Introduction to Reasoning and Proof

Grades 6–8

Denisse R. Thompson
Karren Schultz-Ferrell

The Math Process Standards Series
Susan O'Connell, Series Editor

HEINEMANN
Portsmouth, NH

Heinemann
361 Hanover Street
Portsmouth, NH 03801–3912
www.heinemann.com

Offices and agents throughout the world

Library of Congress Cataloging-in-Publication Data
Thompson, Denisse Rubilee.
    Introduction to reasoning and proof : grades 6–8 / Denisse R. Thompson,
Karren Schultz-Ferrell.
        p. cm.—(Math process standards series)
    Includes bibliographical references.
    ISBN-13: 978-0-325-01733-4
    ISBN-10: 0-325-01733-6
    1. Proof theory—Juvenile literature.   2. Logic, Symbolic and mathematical—
Juvenile literature.   3. Mathematics—Study and teaching (Secondary).
I. Schultz-Ferrell, Karren.   II. Title.
    QA9.54.T46 2008
    510.71'2—dc22   2007044326

Editor: Emily Michie Birch
Production coordinator: Elizabeth Valway
Production service: Matrix Productions Inc.
Cover design: Night & Day Design
Cover photography: Joy Bronston Schackow
Composition: Publishers' Design and Production Services, Inc.
CD production: Nicole Russell
Manufacturing: Louise Richardson

Printed in the United States of America on acid-free paper
12   11   10   09   08   ML   1   2   3   4   5

The authors dedicate this work to our families,
whose love sustains us,
and to our middle school friends and colleagues.

To my parents
—DRT

To Nick, Morgan, and my parents
—KSF

On the CD-ROM

In order to be effective mathematicians, students need to develop understanding of critical math content. They need to understand number and operations, algebra, measurement, geometry, and data analysis and probability. Through continued study of these content domains, students gain a comprehensive understanding of mathematics as a subject with varied and interconnected concepts. As math teachers, we attempt to provide students with exposure to, exploration in, and reflection about the many skills and concepts that make up the study of mathematics.

Even with a deep understanding of math content, however, students may lack important skills that can assist them in their development as effective mathematicians. Along with content knowledge, students need an understanding of the processes used by mathematicians. They must learn to problem solve, communicate their ideas, reason through math situations, prove their conjectures, make connections between and among math concepts, and represent their mathematical thinking. Development of content alone does not provide students with the means to explore, express, or apply that content. As we strive to develop effective mathematicians, we are challenged to develop both students' content understanding and process skills.

The National Council of Teachers of Mathematics (2000) has outlined critical content and process standards in its *Principles and Standards for School Mathematics* document. These standards have become the roadmap for the development of textbooks, curriculum materials, and student assessments. These standards have provided a framework for thinking about what needs to be taught in math classrooms and how various skills and concepts can be blended together to create a seamless math curriculum. The first five standards outline content standards and expectations related to number and operations, algebra, geometry, measurement, and data analysis and probability. The second five standards outline the process goals of problem solving, reasoning and proof, communication, connections, and representation. A strong understanding of these standards empowers teachers to identify and select activities within their curricula to produce powerful learning. The standards provide a vision for what teachers hope their students will achieve.

This book is a part of a vital series designed to assist teachers in understanding the NCTM Process Standards and the ways in which they impact and guide student learning. An additional goal of this series is to provide practical ideas to support teachers as they ensure that the acquisition of process skills has a critical place in their math instruction. Through this series, teachers will gain an understanding of each process standard as well as gather ideas for bringing that standard to life within their math classrooms. It offers practical ideas for lesson development, implementation, and assessment that work with any curriculum. Each book in the series focuses on a critical process skill in a highlighted grade band and all books are designed to encourage reflection about teaching and learning. The series also highlights the interconnected nature of the process and content standards by showing correlations between them and showcasing activities that address multiple standards.

Students who develop an understanding of content skills and cultivate the process skills that allow them to apply that content understanding become effective mathematicians. Our goal as teachers is to support and guide students as they develop both their content knowledge and their process skills, so they are able to continue to expand and refine their understanding of mathematics. This series is a guide for math educators who aspire to teach students more than math content. It is a guide to assist teachers in understanding and teaching the critical processes through which students learn and make sense of mathematics.

Susan O'Connell
Series Editor

This book and the accompanying CD could not have been completed without the encouragement and support of many people. The authors wish to express deep appreciation to all of them.

We extend a special thanks to Sue O'Connell, who saw the vision for this project and initiated our participation, and to Emily Birch, who read and edited this work and advised and encouraged us. We truly appreciate their patience when we ran a bit past deadline.

Thanks to Deborah Heeres, who tried activities with her middle grades students and gave us helpful feedback. To the students who happily shared their thinking while reasoning about increasingly difficult problems, we extend our most grateful appreciation. This work could not have been completed without their enthusiastic cooperation. We also extend a heartfelt thanks to their parents, who gave us permission to use their pictures and work. We particularly thank Alexander, Amy, Drew, and Gavin, who worked additional problems for us during their summer vacation.

Problem-Solving Standard

Instructional programs from prekindergarten through grade 12 should enable all students to—

■ build new mathematical knowledge through problem solving;

■ solve problems that arise in mathematics and in other contexts;

■ apply and adapt a variety of appropriate strategies to solve problems;

■ monitor and reflect on the process of mathematical problem solving.

Reasoning and Proof Standard

Instructional programs from prekindergarten through grade 12 should enable all students to—

■ recognize reasoning and proof as fundamental aspects of mathematics;

■ make and investigate mathematical conjectures;

■ develop and evaluate mathematical arguments and proofs;

■ select and use various types of reasoning and methods of proof.

---

* Standards are listed with the permission of the National Council of Teachers of Mathematics (NCTM). NCTM does not endorse the content or validity of these alignments.

Communication Standard

Instructional programs from prekindergarten through grade 12 should enable all students to—

- organize and consolidate their mathematical thinking through communication;

- communicate their mathematical thinking coherently and clearly to peers, teachers, and others;

- analyze and evaluate the mathematical thinking and strategies of others;

- use the language of mathematics to express mathematical ideas precisely.

Connections Standard

Instructional programs from prekindergarten through grade 12 should enable all students to—

- recognize and use connections among mathematical ideas;

- understand how mathematical ideas interconnect and build on one another to produce a coherent whole;

- recognize and apply mathematics in contexts outside of mathematics.

Representation Standard

Instructional programs from prekindergarten through grade 12 should enable all students to—

- create and use representations to organize, record, and communicate mathematical ideas;

- select, apply, and translate among mathematical representations to solve problems;

- use representations to model and interpret physical, social, and mathematical phenomena.

NCTM Content Standards and Expectations for Grades 6–8

NUMBER AND OPERATIONS

	Expectations
Instructional programs from prekindergarten through grade 12 should enable all students to—	**In grades 6–8 all students should—**
Understand numbers, ways of representing numbers, relationships among numbers, and number systems	• work flexibly with fractions, decimals, and percents to solve problems; • compare and order fractions, decimals, and percents efficiently and find their approximate locations on a number line; • develop meaning for percents greater than 100 and less than 1; • understand and use ratios and proportions to represent quantitative relationships; • develop an understanding of large numbers and recognize and appropriately use exponential, scientific, and calculator notation; • use factors, multiples, prime factorization, and relatively prime numbers to solve problems; • develop meaning for integers and represent and compare quantities with them.
Understand meanings of operations and how they relate to one another	• understand the meaning and effects of arithmetic operations with fractions, decimals, and integers; • use the associative and commutative properties of addition and multiplication and the distributive property of multiplication over addition to simplify computations with integers, fractions, and decimals; • understand and use the inverse relationships of addition and subtraction, multiplication and division, and squaring and finding square roots to simplify computations and solve problems

	Expectations
Instructional programs from prekindergarten through grade 12 should enable all students to—	**In grades 6–8 all students should—**
Compute fluently and make reasonable estimates	• select appropriate methods and tools for computing with fractions and decimals from among mental computation, estimation, calculators or computers, and paper and pencil, depending on the situation, and apply the selected methods;
	• develop and analyze algorithms for computing with fractions, decimals, and integers and develop fluency in their use;
	• develop and use strategies to estimate the results of rational-number computations and judge the reasonableness of the results;
	• develop, analyze, and explain methods for solving problems involving proportions, such as scaling and finding equivalent ratios.

ALGEBRA

	Expectations
Instructional programs from prekindergarten through grade 12 should enable all students to—	**In grades 6–8 all students should—**
Understand patterns, relations, and functions	• represent, analyze, and generalize a variety of patterns with tables, graphs, words, and, when possible, symbolic rules;
	• relate and compare different forms of representation for a relationship;
	• identify functions as linear or nonlinear and contrast their properties from tables, graphs, or equations.
Represent and analyze mathematical situations and structures using algebraic symbols	• develop an initial conceptual understanding of different uses of variables;

	Expectations
Instructional programs from prekindergarten through grade 12 should enable all students to—	**In grades 6–8 all students should—**
	• explore relationships between symbolic expressions and graphs of lines, paying particular attention to the meaning of intercept and slope;
	• use symbolic algebra to represent situations and to solve problems, especially those that involve linear relationships;
	• recognize and generate equivalent forms for simple algebraic expressions and solve linear equations
Use mathematical models to represent and understand quantitative relationships	• model and solve contextualized problems using various representations, such as graphs, tables, and equations.
Analyze change in various contexts	• use graphs to analyze the nature of changes in quantities in linear relationships.

GEOMETRY

	Expectations
Instructional programs from prekindergarten through grade 12 should enable all students to—	**In grades 6–8 all students should—**
Analyze characteristics and properties of two- and three-dimensional geometric shapes and develop mathematical arguments about geometric relationships	• precisely describe, classify, and understand relationships among types of two- and three-dimensional objects using their defining properties;
	• understand relationships among the angles, side lengths, perimeters, areas, and volumes of similar objects;
	• create and critique inductive and deductive arguments concerning geometric ideas and relationships, such as congruence, similarity, and the Pythagorean relationship.

	Expectations
Instructional programs from prekindergarten through grade 12 should enable all students to—	**In grades 6–8 all students should—**
Specify locations and describe spatial relationships using coordinate geometry and other representational systems	• use coordinate geometry to represent and examine the properties of geometric shapes; • use coordinate geometry to examine special geometric shapes, such as regular polygons or those with pairs of parallel or perpendicular sides.
Apply transformations and use symmetry to analyze mathematical situations	• describe sizes, positions, and orientations of shapes under informal transformations such as flips, turns, slides, and scaling; • examine the congruence, similarity, and line or rotational symmetry of objects using transformations.
Use visualization, spatial reasoning, and geometric modeling to solve problems	• draw geometric objects with specified properties, such as side lengths or angle measures; • use two-dimensional representations of three-dimensional objects to visualize and solve problems such as those involving surface area and volume; • use visual tools such as networks to represent and solve problems; • use geometric models to represent and explain numerical and algebraic relationships; • recognize and apply geometric ideas and relationships in areas outside the mathematics classroom, such as art, science, and everyday life.

MEASUREMENT

	Expectations
Instructional programs from prekindergarten through grade 12 should enable all students to—	**In grades 6–8 all students should—**
Understand measurable attributes of objects and the units, systems, and processes of measurement	• understand both metric and customary systems of measurement; • understand relationships among units and convert from one unit to another within the same system; • understand, select, and use units of appropriate size and type to measure angles, perimeter, area, surface area, and volume.
Apply appropriate techniques, tools, and formulas to determine measurements	• use common benchmarks to select appropriate methods for estimating measurements; • select and apply techniques and tools to accurately find length, area, volume, and angle measures to appropriate levels of precision; • develop and use formulas to determine the circumference of circles and the area of triangles, parallelograms, trapezoids, and circles and develop strategies to find the area of more-complex shapes; • develop strategies to determine the surface area and volume of selected prisms, pyramids, and cylinders; • solve problems involving scale factors, using ratio and proportion; • solve simple problems involving rates and derived measurements for such attributes as velocity and density.

DATA ANALYSIS AND PROBABILITY

Instructional programs from prekindergarten through grade 12 should enable all students to—	Expectations
	In grades 6–8 all students should—
Formulate questions that can be addressed with data and collect, organize, and display relevant data to answer them	• formulate questions, design studies, and collect data about a characteristic shared by two populations or different characteristics within one population;
	• select, create, and use appropriate graphical representations of data, including histograms, box plots, and scatterplots.
Select and use appropriate statistical methods to analyze data	• find, use, and interpret measures of center and spread, including mean and interquartile range;
	• discuss and understand the correspondence between data sets and their graphical representations, especially histograms, stem-and-leaf plots, box plots, and scatterplots.
Develop and evaluate inferences and predictions that are based on data	• use observations about differences between two or more samples to make conjectures about the populations from which the samples were taken;
	• make conjectures about possible relationships between two characteristics of a sample on the basis of scatterplots of the data and approximate lines of fit;
	• use conjectures to formulate new questions and plan new studies to answer them.
Understand and apply basic concepts of probability	• understand and use appropriate terminology to describe complementary and mutually exclusive events;
	• use proportionality and a basic understanding of probability to make and test conjectures about the results of experiments and simulations;
	• compute probabilities for simple compound events, using such methods as organized lists, tree diagrams, and area models.

The Reasoning and Proof Standard

The ability to reason systematically and carefully develops when students are encouraged to make conjectures, are given time to search for evidence to prove or disprove them, and are expected to explain and justify their ideas.

—National Council of Teachers of Mathematics,
Principles and Standards for School Mathematics

Why Focus on Reasoning and Proof?

The mathematics classroom of the past might have been one of the quietest places in the school. Opportunities for students to explain thinking and reasoning were rarely offered. In today's classrooms, however, reasoning is viewed as a necessary process to ensure that students understand mathematics concepts and skills. "Many mathematicians consider the NCTM standard concerning reasoning and proof to be its most important" (Chapin, O'Connor, and Canavan Anderson 2003, 78). Students benefit when they have opportunities to explain thinking and reasoning not only through discourse but also through recording of representations (e.g., charts, graphs, drawings, diagrams) of mathematical thinking. Students should also be given opportunities to reflect on their thinking and reasoning through writing. These experiences help students extend their thinking, solidify understandings about concepts and skills, and learn the different ways their classmates think about, reason with, and solve mathematics problems and situations. Some argue that "students cannot understand mathematics unless they acquire and employ the forms of argument used to justify mathematical propositions, and they cannot appreciate the power of these forms of thinking unless they are deployed in learning important mathematical ideas" (Valentine, Carpenter, and Pligge 2005, p. 96).

Although we realize the importance of questioning students about their thinking when errors and misconceptions are evident, we often fail to question them when their solutions are correct. As a result, many students become conditioned to assume that any questions about their work imply something is wrong with it. But listening to students as they explain their reasoning, whether correct or incorrect, gives us valuable information about what they know and are able to do. Students' strengths and misconceptions are revealed, which provides information that helps us plan instruction to meet the needs of all students. In addition to helping us gain information about students, our questioning delivers important messages to our students: their ideas are valued and important; mathematics is less about memorization and more about reasoning; and what they are learning should make sense (Burns 1997).

As previously noted, reasoning can be expressed through representations as well as through oral justifications. These representations become evidence of learning that can be shared later with peers. When students become confident in explaining and representing thinking and reasoning, they are able to write about their reasoning. When we emphasize reasoning and proof in our classrooms, our students become engaged in different types of reasoning and begin to understand how to express acceptable mathematical explanations.

What Is the Reasoning and Proof Standard?

The National Council of Teachers of Mathematics (NCTM) has recommended standards that can be used as a resource and a guide for teachers as they plan and create instructional lessons and activities to develop students' understandings about mathematics. The first five standards focus on mathematical content: number and operations, algebra, geometry, measurement, and data analysis and probability. The next five standards address the processes by which students explore and use mathematics as well as develop understandings about mathematics concepts and skills: problem solving, reasoning and proof, communication, connections, and representation. The process standards should be embedded throughout our mathematics instructional program.

Instructional programs from prekindergarten through grade 12 should enable all students to—

■ recognize reasoning and proof as fundamental aspects of mathematics;

■ make and investigate mathematical conjectures;

■ develop and evaluate mathematical arguments and proof; and

■ select and use various types of reasoning and methods of proof. (NCTM 2000, 56)

This book explores ways to assist teachers as they help students extend their thinking, deepen understandings, and make sense of mathematics through the process standard of reasoning and proof.

Developing Skills and Attitudes

"Reasoning mathematically is a habit of mind, and like all habits, it must be developed through consistent use in many contexts" (NCTM 2000, 56). The ability to reason develops over time with multiple experiences. We must provide our students more than one reasoning opportunity a week. Reasoning must become a part of the way we teach daily in order for our students to become proficient in their ability to reason.

Establishing a classroom environment that is both respectful and supportive is the first step in helping our students become comfortable enough to offer explanations of mathematical thinking and reasoning with confidence. Ground rules are needed to create a classroom climate in which students listen respectfully to one another during mathematics discussions. Students must feel safe when expressing their thinking and sharing their own ideas for examination. They will not contribute to a discussion if they feel their ideas will be ridiculed or dismissed quickly. Ground rules help with this, and we must be consistent with our expectations of behaviors during these discussions.

As teachers build a supportive environment, they must set the expectation that all students will listen to what others say and participate actively in an ongoing discussion. Students need our support to develop the skills necessary to participate in teacher-to-student, student-to-student, and student-to-teacher discourse. They must have daily opportunities to talk and reason about mathematics if we want them to be able to represent and write about mathematical reasoning.

Presenting tasks to students that are open-ended and rich in context provides them with opportunities to develop reasoning skills. When we present problems to our students that can be solved in more than one way, or that have more than one correct answer, we promote talk about thinking and reasoning. Problems that cause students to stretch their thinking are a must. This book addresses the considerations introduced in this section, as well as others, to help you begin to implement the reasoning and proof standard into your daily instruction.

How This Book Will Help You

This book is designed to help you better understand the NCTM Process Standard of reasoning and proof for students in grades 6 through 8. It explains how these students' reasoning abilities differ from what they were capable of in the elementary grades and examines how students' reasoning experiences in grades 6 through 8 develop understandings necessary for making formal proofs in later school years. Specific types of reasoning are described, and student dialogues are provided throughout the book to model how students use these different ways to reason. The teachers facilitating these student dialogues model critical questioning that promotes rich discussions in which students reason about the mathematics they are learning.

We share strategies that teachers can use to support students as they develop reasoning skills, and the CD provides various activities that promote reasoning. We discuss how you can provide reasoning experiences for your students in which they will make conjectures about important mathematical ideas and relationships. In grades 6 through 8, they should be expected to support their conjectures by developing and evaluating mathematics arguments. Student dialogues are provided to demonstrate how students in the middle grades begin to make conjectures and then justify their reasoning with informal or formal mathematical arguments.

The activities provided on the CD do not list specific grade levels because the activities have been formatted so that you can modify them to meet the needs, interests, and skills of your students. We describe how to do this in the "About the CD-ROM" section near the end of this book. We encourage you to choose activities that fit your content and instructional goals. The problems and activities illustrated for you in the student dialogues throughout the book are included on the CD. You will find these activities to be both engaging and challenging for your students, and facilitating discussions with your students about the problem solving within the activities will support their reasoning skills. These discussions can occur either while your students are solving the tasks as a whole group or when you talk with them about their problem solving and reasoning after they have completed an activity. In addition to the CD activities, you will find a listing of resources that will help your students make sense of mathematics, develop their reasoning ability, and strengthen their grasp of mathematics concepts and skills. Also included on the CD are rubrics that will help you assess your students' development of reasoning skills, including rubrics you can use to monitor your students' development of reasoning skills or explanations of their thinking when solving a problem, and a rubric students can use to monitor their own reasoning progress.

As mentioned earlier, examples of student discourse demonstrate how students reason about important mathematical ideas and relationships and serve as a model for how to facilitate discussions that promote reasoning in your own classroom. These examples demonstrate how students reason about a variety of mathematical concepts and skills. In addition, examples of student work are presented that provide another glimpse into students' thinking as their reasoning and proof skills develop. Practical tips and ideas are shared in the "Classroom-Tested Tip" boxes to help you implement the ideas explored in the chapters. Following each chapter, you will find several discussion questions to guide your reflection on the content of the chapter, either alone or with a group of your colleagues.

Chapter 5 discusses how the process standard of reasoning and proof is supported by the remaining process standards: problem solving, communication, representation, and connections. We discuss how students' reasoning helps them make vital connections to their everyday lives and other content areas in mathematics. We share ideas for how to provide students with meaningful tasks that connect reasoning to all the process standards. In Chapter 6, a variety of ways to assess students' reasoning skills are examined. After we have explored the reasoning and proof standard in depth, you will see how it relates to the different content areas in mathematics in Chapter 7, "Reasoning and Proof Across the Content Standards."

The role of communication in supporting students' reasoning has been stressed several times in this chapter, especially communicating reasoning through discourse. It is important to remember that we cannot expect our students to write about mathematical reasoning if they have not talked about it first. In fact, some educators consider oral sharing of reasoning to be a "stepping-stone on the path to convincing others with written argument" (Boats, Dwyer, Laing, and Fratella 2003, p. 211). Students' ability to communicate their thinking and reasoning orally as well as through writing and pictorial and symbolic representations is examined in various chapters, and examples of student work are included throughout the book.

We hope this book will enhance your understanding of the reasoning and proof standard and provide you with practical ideas and strategies that will help you to facilitate your students' development of reasoning and proof skills. When we grow in our understandings about reasoning and proof, so will our students.

Creating rich mathematical tasks and asking questions related to justification encourage students to examine what constitutes a valid justification for a general statement. Such reasoning bridges the mathematical activity that occurs at the elementary level and the reasoning that is necessary at the high school level. (Lannin, Barker, and Townsend 2006, p. 443).

Questions for Discussion

1. To what extent did mathematics make sense to you when you were in school? How often were you asked to explain how you solved a problem? How might your past experiences and attitudes about reasoning affect how you think students are able to reason?

2. If students state the correct answer but cannot explain why the answer makes sense mathematically, what questions might you have about your students' understanding? How might this inability to explain and reason about mathematics present difficulties later in school?

3. What attitudes or dispositions are crucial for explaining one's thinking and reasoning? How does the atmosphere in the classroom affect students' abilities to develop reasoning?

4. Reflect on the quote included at the end of this chapter by yourself or with colleagues. Compare students who learn mathematics by reasoning with students who learn mathematics only through procedures.

CHAPTER 1

Creating a Learning Environment That Promotes Reasoning and Proof

Mathematical reasoning develops in classrooms where students are encouraged to put forth their own ideas for examination.

—National Council of Teachers of Mathematics,
Principles and Standards for School Mathematics

Our Attitude Toward Mathematics

Perhaps the most important factor in creating a classroom atmosphere that encourages thinking and reasoning is our attitude toward mathematics. "How a teacher views mathematics and its learning affects that teacher's teaching practice, which ultimately affects not only what the students learn but how they view themselves as mathematics learners" (National Research Council 2001, 132). Grades 6 through 8 are often the time when many students begin to dislike mathematics. Indeed, although 73 percent and 67 percent of male and female fourth-grade students, respectively, reported liking mathematics on the 2000 National Assessment of Educational Progress, these percentages dropped to 56 percent and 51 percent, respectively, at the eighth-grade level (Lubienski, McGraw, and Strutchens 2004). So, teachers' enthusiasm for mathematics is essential. When teachers are excited about mathematics, they encourage their middle grades students to exhibit that same enthusiasm.

   As your students begin to share their thinking and reasoning, you will likely discover they have a diverse range of reasoning abilities. Many of the problems used to encourage reasoning are non-routine problems for which there is no algorithm. Even when solving traditional problems, students may resort to unusual approaches as they reason about a problem when they do not understand how to complete an algorithm to solve it. When students attempt to make sense of mathematics in ways that are not traditionally taught, their reasoning should be respected. As we honor students' unique approaches to reasoning about mathematics, they come to realize that their creative and critical thinking is important and valued. We must instill confidence in students so they are ready for the challenges we offer them—challenges that will develop and strengthen their reasoning abilities. In creating a classroom that is a mathematics community, we must show our students that we value mathematics, that we want it to make sense to them, and that reasoning is central to everything we do.

A Safe, Supportive, and Respectful Classroom

In order for students to be comfortable in communicating their mathematical thinking and reasoning, a supportive learning environment must be in place. A student's classroom can be described as a home away from home, so it is natural that we would want to create a classroom atmosphere that provides support and respect for all students. If students feel their ideas will be criticized or ridiculed, they will not be willing to share how they are thinking and reasoning about mathematics. To create a classroom environment that promotes reasoning, we must first establish expectations that ensure a supportive environment in which our students can talk confidently about their thinking and reasoning. These expectations must be clearly understood by all students and consistently reinforced throughout the school year.

It is equally important that we establish a respectful classroom environment. We want our students to know that everyone in the classroom has a right to be heard and that everyone's thinking and reasoning will be considered in a respectful manner. Along with this right for ideas to be heard and considered, we want our students to know there is an expectation that everyone will listen to others' thinking and reasoning. Listening to their classmates carefully will help students understand one another's reasoning and ideas. When students are actively engaged in listening, it enables them to participate more fully in ongoing discussions. As your students begin to share their reasoning, expect that there will be students who disagree with others' thinking. When disagreements occur about how a student is reasoning, the challenge must be accompanied by an explanation that is offered respectfully. The establishment of a safe, supportive, and respectful classroom environment is the first step in setting the stage for reasoning and proof.

CLASSROOM-TESTED TIP

You might consider posting the following suggestions for students to help them become better listeners; they can refer to the suggestions throughout the year as needed.

■ Give all your attention to the student who is speaking.

■ Change how you are sitting so you can refocus on the speaker if you are beginning to daydream.

- Wait until the speaker has finished talking before you begin to talk. Be courteous.

- Listen for the important ideas the speaker is sharing so you can speak to those ideas.

- Give yourself time to finish listening to the speaker before you begin talking. Then you can begin to think about what you will say in response to the speaker or about another point you would like to discuss about the mathematics.

- Ask questions of the speaker if you do not understand something the speaker said.

- Repeat a student's reasoning in your words to help you understand after the speaker has shared his reasoning.

Engaging and Challenging Mathematics

What else can we do to support our students as they develop reasoning skills? We must choose appropriate tasks that are engaging and thought-provoking for our students—problems that are rich in context, that are meaningful and relate to students' everyday lives, and that are challenging experiences. Consider the following *Containers of Nuts* task (also see the task on the CD):

> Cassandra must distribute 12 pounds of nuts among three identical containers according to the following rules:
>
> - Each container has a whole number of pounds.
>
> - Each container has some nuts in it.
>
> - No two containers can have the same number of pounds.
>
> - All 12 pounds of nuts must be used.
>
>   Find all the possible ways Cassandra can distribute the nuts among the three containers. Explain how you know you have found all the different ways.

Not only does this problem elicit multiple solutions from students, but it also requires students to think logically about the mathematics in order to find all the possible combinations for the containers and to explain how they know all the solutions have been found. Note that the mathematics in this problem should be straightforward and accessible to middle grades students, regardless of prior knowledge or computational facility. The focus is on reasoning and expecting students to discuss their thinking; it is not on computation, so using numbers for which computation is not a hindrance is important in order to engage all students in the problem.

In solving rich, or nonroutine, problems, students use a variety of problem-solving strategies, such as making an organized list, completing a chart, and using representations. Students will also use reasoning when they explain and justify how they know they have identified all of the possible combinations.

Middle grades students grappled with this container problem. As they worked, the teacher facilitated a discussion about the problem. Please note that in the following dialogue, different students responded throughout the discussion although student names are not indicated. This is the organization used in all the student dialogues in this book.

TEACHER: I see that several of you have started building a table to show the number of pounds of nuts in each container. Let's compare solutions. Does someone want to share a solution and explain why it works?

STUDENT: I found 6 pounds for the first container, 5 pounds for the second container, and 1 pound for the third container. The three amounts add to 12 pounds, and they are all whole numbers.

TEACHER: Are there any other solutions with 6 pounds in one of the containers?

STUDENT: I found 6, 3, and 3. Oh, wait. That doesn't work. They add up to 12, but you can't have the same amount in two containers. So, 6, 3, 3 can't be one of the ways.

TEACHER: That's a good observation. You also demonstrated an important thing to remember when working on problems—always go back and check to see if a solution makes sense for the problem. In this case, for an arrangement to satisfy the problem, it has to meet all the conditions of the problem, and one of those is that no two containers have the same amount. What other solutions do we have?

STUDENT: I found 9 pounds for the first container, 2 pounds for the second container, and 1 pound for the third container. The total is 12 pounds, all the pounds are whole numbers, and I didn't use the same number twice. So, it works.

STUDENT: Is 9 pounds, 2 pounds, and 1 pound the same arrangement as 1 pound, 2 pounds, and 9 pounds?

TEACHER: What do you think? (*Students raise their hands.*)

STUDENT: The problem says to find the different ways for the nuts, so I think 9, 2, and 1 is the same as 1, 2, and 9. Our group just listed this combination one time.

STUDENT: We wondered about that too, but we decided to list both ways. But it would be easier to just list it one time; then we won't count it twice.

TEACHER: Could we do the problem either way? (*Teacher waits a few seconds for any other comments, but there are none.*) Sometimes the order of an arrangement does matter. For instance, in a race it matters who comes in first, second, or third place. If we had three containers of different shapes, or if they were presents for different people, it might matter who got 9 pounds, 1 pound, or 2 pounds. But for now, let's just think of them as the same arrangement. How many different arrangements have you found?

STUDENT: We found 7.

STUDENT: We found 5. Did you count the same way twice? (*directed to the previous student*)

TEACHER: We have two different solutions here for the total number of arrangements. How can we decide if we have found all of them?

STUDENT: (*from the group that found 7 ways*) We tried to find more, but we couldn't. Any other ways used the same number for two containers, and that's not allowed.

STUDENT: (*from another group that found 7 ways*) I went through all the numbers from 1 to 12 and couldn't find any more combinations that worked.

STUDENT: (*from the group that found 5 ways*) We checked our answers with another group. We missed two of the ways that use 1 pound.

TEACHER: So, how can you know that you have found all the ways and didn't miss any? (*Teacher waits for a few seconds.*) Let's consider a specific case. How can you know that you have found all the possibilities with one pound in a container?

STUDENT: We could list 1 pound and then 2 pounds (*with 9 pounds in the third container*), 1 pound and 3 pounds (*with 8 pounds in the third container*), 1 pound and 4 pounds (*with 7 pounds in the third container*), 1 pound and 5 pounds (*with 6 pounds in the third container*). When we try to list 1 pound and 6 pounds (*with 5 pounds in the third container*), it's already on the list from 1 pound and 5 pounds. Then do the same thing starting with 2 pounds and 3 pounds.

TEACHER: What do the rest of you think about her reasoning? (*Students nod in agreement.*) Let's have this group put their table on the board. (See Figures 1–1 and 1–2.) What patterns do you see in the table?

Container 1	Container 2	Container 3
1 lbs	2 lbs	9 lbs
1 lbs	3 lbs	8 lbs
1 lbs	4 lbs	7 lbs
1 lbs	5 lbs	6 lbs
2 lbs	3 lbs	7 lbs
2 lbs	4 lbs	6 lbs
3 lbs	4 lbs	5 lbs

Figure 1–1 *A student's strategy for organizing the possible arrangements of nuts in three containers.*

Figure 1–2 *A group of students discussing the Containers of Nuts problem at the
board.*

STUDENT: The first column stays 1 for a long time while the second column goes up
by 1 and the third column goes down by 1. When the first column is 1, the numbers in the next two columns have to add to 11. You keep doing this until the
numbers repeat. Then you make the first column 2 and do the same thing, and
then 3, and so on. But the first column can't be bigger than 3 because the numbers start to repeat.

TEACHER: I would like you first to think about what we've been discussing and then
write how you were thinking about the problem. A written record can be helpful
when we have another problem that is similar to this one. (*Students were given
an opportunity to record their thought processes in a math journal.*)

This simple problem was challenging for students, not in terms of finding arrangements for the containers but in justifying how they knew all the arrangements had
been found. Students used an organized list and patterns within the list to solve the
problem and justify how they knew they had all the possible combinations. Thus, the
problem provided an opportunity to blend an important problem-solving strategy
with reasoning skills. In addition, the discussion about whether different arrangements of the same numbers should be considered as the same or a different combination sets the stage for later work with permutations. When this topic occurs in the
curriculum, the teacher can revisit the *Containers of Nuts* problem and students can
reason that each arrangement of three numbers generates six separate rows in the
table when order is important.

The teacher implemented important strategies in presenting the problem, guiding the problem-solving process, and facilitating the resulting student discussion:

■ She made a copy of the problem available to students throughout the problem-solving process. Teachers can make the problem available in various ways: each student can have a copy; a group can share a copy; a copy can be displayed on an overhead projector or displayed on a white board. Having the problem visible throughout the problem-solving process helps students focus their thinking.

■ She used a problem that had multiple solutions.

■ She expected students to explain their problem-solving strategies and why their solutions made sense. In the process, she helped students focus on the constraints in the problem.

■ She gave students wait time during the student discussion. Having some time to reflect allowed students to think about the question being asked and formulate their response.

■ She asked probing questions to help students clarify their thinking and reasoning.

■ She had students work collaboratively to solve the problem.

■ She helped students make connections to similar ideas and discover generalizations.

■ She encouraged students to record their solutions in writing, which helped them reflect on the problem during and after solving it.

The containers problem can be easily adapted, extended, and differentiated by changing the total number of pounds or the constraints. If the total number of pounds is larger, more arrangements are possible and students have further opportunities to investigate the use of patterning to solve and reason about the problem.

A natural extension is to remove one of the constraints and ask:

How will your answer change if two containers can have the same number of pounds in them?

When the students from the earlier dialogue were given this extension after writing about their reasoning on the problem, they recognized that additional possibilities would be added to the table. One student noted that she would "have 5 more options; doubling the numbers 1 to 5," meaning she would have (1, 1, 10 pounds; 2, 2, 8 pounds; 3, 3, 6 pounds; 4, 4, 4, pounds; and 5, 5, 2 pounds).

A second adaptation could simplify the problem to two containers but allow connections to be made to algebra and graphing:

Cassandra must distribute 12 pounds of nuts among two containers of different shapes. Find all the possible ways Cassandra can distribute the nuts among the two containers if each container must hold a whole number of pounds.

Because the two containers have different shapes, it is natural to assume an arrangement of 3 pounds in the first container and 9 pounds in the second container is different from 9 pounds in the first container and 3 pounds in the second container. So, if *f* represents the number of pounds in the first container and *s* the number of pounds in the second container, then this situation can be described by the equation $f + s = 12$. Although students could reason about the problem in the same way as before, they could now also make connections to algebra. They can make a table of values showing the possible combinations for *f* and *s*; when students graph these values, they obtain a set of discrete points on the line in the first quadrant. If a container can hold 0 pounds, then the two intercepts for the graph are also included. (See Figure 1–3.) So, depending on the goals for the class, a problem used to engage students' reasoning abilities can also permit entry into important concepts in algebra.

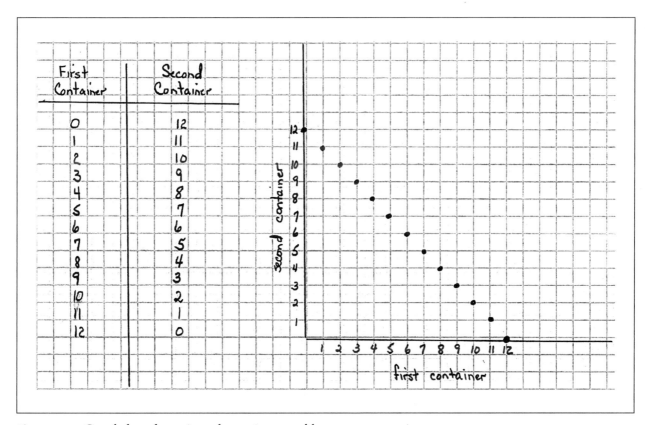

Figure 1–3 *Graph for adaptation of containers problem to two containers.*

CLASSROOM-TESTED TIP

Choosing tasks that engage students in reasoning may at first be difficult. Consider the following when planning your lessons:

- Choose tasks that are of interest to your students, those that come from their environment. An example of an authentic problem for students to solve and reason about is asking them to determine the number of items needed for a class party and which quantities are the better buy.

- Consider tasks that will motivate your students to investigate a mathematical situation. Students' natural curiosity about the world can easily be the driving force in a mathematics investigation that is embedded with reasoning opportunities.

- Choose tasks that encourage multiple perspectives and representations, such as tables, graphs, pictures, or equations. This helps students reason about mathematical ideas and relationships in different ways and enables them to consider others' views.

- Present tasks that are not easily solved but within your students' abilities. This promotes students' persistence in problem solving.

The accompanying CD provides activities and tasks such as the *Containers of Nuts* problem that will actively engage your students in meaningful, challenging mathematics. These activities support students' ability to think deeply about a variety of mathematics concepts and skills as well as develop their reasoning skills. They are beneficial to motivate students to talk meaningfully about the mathematics they are learning. Your facilitation of these discussions is critical to increase students' understanding of mathematics and develop their reasoning abilities.

CLASSROOM-TESTED TIP

Here are several ways to engage students in a discussion:

- Think-pair-share is a strategy that is effective to promote student talk (McTighe and Lyman 1988). Students are first given time to think quietly about a question that has been asked. This time for students to think silently in order to develop and organize their ideas is an important component of the strategy because it enhances students' responses to questions.

Each student then pairs with another student to talk about what he or she was thinking regarding the question. This brief discussion between partners gives students an opportunity to share ideas, clarify each other's thinking, and even challenge and ask questions of one another. Students will be more willing to take risks and share mathematical thinking and reasoning with the class because they have already discussed their thinking with a partner. Students then share ideas with the whole group. They either share what they were thinking or something they discussed with their partner. It is important that they be able to share their partner's thinking and reasoning as well as their own. Students often feel more comfortable talking in a whole group after they have processed a question in this manner.

■ Using index cards—one per student—is a way to ensure that all students have an opportunity to contribute in a student discussion. Record each student's name on a card and draw a card at random when calling on students throughout the discussion. This strategy is effective initially in promoting more student talk from everyone. Once students feel comfortable sharing in daily discussions, it is likely all your students will be eager to share their thinking and reasoning!

Questioning

Another consideration when supporting students as they develop reasoning skills is the types of questions we ask. Students will be denied the opportunity to reason if the only questions we ask are those that merely require one-word answers. However, questions that require students to ponder about mathematical ideas encourage reasoning, and asking good questions facilitates productive learning. Here are several questions or prompts that encourage students to think and reason:

■ How do you know? (Ask this question even if the student gave a correct answer!)

■ What other ways could you solve this problem?

■ Tell why you agree or disagree with the way the problem was solved.

■ What strategy did you use in playing the game? How was this strategy helpful?

■ What is different about these answers?

■ What is another example of _____ ?

■ How are _____ and _____ alike? How are they different?

Although questions that elicit one-word answers are appropriate at times, questions that require students to think and reason about mathematics should be the focus in every lesson. Ask questions that allow students to reflect, require them to make conjectures, or cause them to justify their reasoning. Asking effective questions will promote students' use of critical thinking skills, their ability to establish relationships, and their ability to make meaningful connections. The following questions also help students see that explanations can be supported or refuted by evidence:

■ Why does this work? Will it work every time?

■ If we try this with more problems, will it still work?

■ Does Morgan's explanation make sense to you?

■ What would happen if you . . . ?

■ How will you convince us this is true?

The questions outlined above are similar to those in the Advancing Children's Thinking framework (Fraivillig 2001), which describes teachers' interactions to engage students in productive mathematics discourse. In this framework, teachers *elicit* various solution methods to a problem, expect students to elaborate, and use students' explanations as a springboard for class discussion. Then, teachers *support* students by helping them make connections among similar problems and encouraging them to seek help as needed. Finally, teachers *extend* students' thinking by encouraging and expecting them to make generalizations and find alternative solution approaches.

As you begin to encourage more reasoning and proof in your classroom, explain to students the changes that will be occurring during your mathematics discussions. Students need to know that you will be asking questions to find out how they are thinking about the mathematics they are learning. This helps them to understand your expectations and to realize that talking about their thinking and reasoning will improve their understanding of the mathematics being taught. It is important to keep in mind that sharing thinking and ideas in mathematics may be uncomfortable for many students if they have not had these opportunities in previous grades. Some students may take longer to feel confident enough to do this. It is a gradual process but one that we need to continue until students are comfortable with its use.

Choose one or two questions from the bulleted lists to focus on in the beginning, and then as you and your students become comfortable with those questions, expand on the types of questions you ask. When you model good questioning and facilitate how you want your students to participate in discussions, they will begin to ask the same questions of their classmates and become more comfortable in contributing to discussions. Eventually you want your students to be comfortable asking questions of you and of each other.

Think about these considerations when choosing questions to ask students:

- Choose questions that require students to do more than recite a fact from memory or restate a procedural skill. Questions that cause students to question their own thinking will demand more from them cognitively and will develop stronger mathematical understandings. Asking, "How can you find $\frac{3}{4}$ of $\frac{2}{3}$ in two different ways?" or "What are some situations in which you need to find $\frac{3}{4}$ of $\frac{2}{3}$?", requires students to think more deeply than they would if they were simply asked, "What is $\frac{3}{4}$ of $\frac{2}{3}$?", which generates only a single numerical answer.

- Ask probing questions after students have answered the initial question. These questions stimulate students' thinking and present you with more information about how they are thinking. Probing encourages students to think about their thinking and provides an opportunity for them to uncover any errors in their original reasoning, which will help them further clarify their thinking. Even if a response is correct, a probing question might uncover a misconception a student has about the mathematics involved. Several examples of probing questions are: "What did you find out when you multiplied?", "Will this work every time? Why or why not?", and "What's another example of how that will work?"

- Choose problems or tasks that solicit several or multiple acceptable answers to motivate students in problem solving. The *Containers of Nuts* task in this chapter demonstrates this type of problem.

- Use prompts that support your students' collaboration to make sense of a problem-solving situation. You might say, "Convince all of us why your group's reasoning makes sense."

- Choose questions that help students make connections in mathematics, such as "How do the values in this table relate to a graph?" and "How would understanding _____ help us _____ ?"

- Help students see connections among similar problems. Problems in which a total amount needs to be partitioned into two or more groups are similar to the *Containers of Nuts* problem. The strategy used to solve the containers problem can be used with similar problems as well.

- Model questions that strengthen students' self-monitoring skills. It is important that students know when their answers are correct. Questions such as "How did you know you had the correct solution?" and "Why does that answer make sense to you?" model the self-monitoring process for students.

Additional Factors That Promote Reasoning

Other considerations that help create an environment that promotes thinking and reasoning are briefly discussed in this section. Each of these ideas is discussed in more detail in a later chapter.

Discourse

Make it a priority for your students to talk daily with you and their classmates about the mathematics concepts and skills they are learning, which helps them not only to internalize the mathematics they are learning but also to obtain a deeper understanding of it. A goal is to strive for less teacher talk and more student talk, including student-to-student discussions. Ask questions that generate productive discussions and then become the facilitator to guide these discussions. In addition to talking about the mathematical concepts they are learning, it is important for students to talk about the procedures and strategies they are using to solve problems. One important benefit of discourse is that in order for students to express their thinking in words, they must first reflect on and then clarify their thinking. When students listen to their classmates' thinking and reasoning, it enhances their own learning. An additional benefit of student discussions is that we are presented with multiple opportunities to gather information informally about what our students know and are able to do. These daily student discussions about their thinking and reasoning become ongoing assessment that is an integral part of our classroom instruction.

CLASSROOM-TESTED TIP

Incorrect or Flawed Reasoning

Incorrect or flawed reasoning will occur when your students begin to contribute their thinking more in daily lessons. We cannot expect that students' reasoning will be flawless. As students routinely begin to reason about the mathematics they are learning, they must understand that it is acceptable to reflect on, analyze, and rethink others' reasoning, especially reasoning that is flawed. All students should know what it is like to think through reasoning that does not work or make sense.

Treat these moments as learning opportunities for both you and your students. The flaws in a student's reasoning often reveal misconceptions that could be shared by other students in the group. It is beneficial to allow students time to think through flawed reasoning. Convey to students that making errors because of misunderstandings is part of the reasoning process, and that these misconceptions will be examined with the intent of learning from them. Common student errors can themselves become tasks in which students evaluate others' reasoning to determine its validity; several activities using this technique are included on the CD.

Ask additional questions to probe for clarification, which helps students examine their thinking more closely and allows other students the opportunity to analyze the flaws in their classmates' reasoning. After considering their reasoning more thoroughly, students should be given the opportunity to revise flawed reasoning about problem solving, incorrect conjectures, or incomplete mathematical arguments.

Recording Sheets

When students record how they are thinking and reasoning in solving a problem, it also helps them organize their thinking. A recording sheet allows students to reflect on the mathematics completed, and it provides them with evidence that is needed to formulate a conjecture and/or develop and justify mathematical arguments. Many of the activities contained on the CD encourage students to record representations of their work as they solve the tasks. These activities also ask students to record their thinking and reasoning about the mathematics in the task.

Writing

When students write about mathematics, it helps them make sense of what they are learning. They must organize their thinking in order to record these understandings and reflections. Students' written responses also provide us with information we can use to assess their understanding (Countryman 1992). The types of writing products vary; the most popular are journals or logs. Writing is an excellent vehicle for students to explain their reasoning, and it can be used for different purposes and formats. The feedback we provide to students about their writing products is important. Providing informational feedback that is specific, supportive, and honest is critical if we want students to develop clear mathematics understandings. This is especially important when a student's reasoning does not make sense or is flawed. Feedback that enables a student to reflect on his thinking is far more beneficial than a message that simply states, "Almost there!"

CLASSROOM-TESTED TIP

Below are several ways to incorporate reasoning in students' writing activities:

- After completing an activity, students write to propose a conjecture or a generalization.

- At the conclusion of a mathematics lesson, students write a letter to an absent student to explain the reasoning the individual student or the group used in the day's mathematics task.

■ Students write to explain reasoning in describing a mathematical relationship, such as how the graph connects to the values in a table.

■ Students write to explain their understanding of a concept, such as division of fractions.

Wait Time

Students need time to organize their thinking and to formulate an explanation of their solution before answering a question. Every student deserves the same amount of wait time. Wait ten to twenty seconds after asking a question of the whole class. When you call on a specific student, give that student additional time to organize his or her thinking before expecting a response. This gives a student time to think through a question and time to express his or her thinking and reasoning in words. Wait time is a beneficial expectation for us as well! After a student has responded to a question, give yourself ten seconds to think about the student's answer and how you will respond to it. This time allows us to think about any follow-up or probing questions that will continue to guide a productive discussion.

Probing Questions

Ask questions that encourage students to add more detail to explanations, thus revealing more of what they understand about the mathematics they are learning. Once students have observed you modeling probing questions, encourage them to ask questions about other students' thinking. This supports them in further developing understandings. The Classroom-Tested Tip on page 17 provides additional information about probing questions, and examples of how to implement this type of questioning are modeled in the student dialogues throughout the book. Other examples of questioning techniques can be found in the articles listed in the Resources section at the end of the book (see, for example, Manouchehri and Lapp 2003).

Group Work

Provide daily opportunities for students to work in pairs, trios, or small groups to solve and discuss problems. Use both homogeneous and heterogeneous groups when planning small-group work within a large-group lesson. More than likely many of these small groupings will happen spontaneously as the need to discuss a problem-solving task occurs. Small-group work allows students to discuss problems together in a less threatening environment and to compare their ideas with those of others. These experiences give them opportunities to adjust or consolidate their thinking. Group work also provides students opportunities to discuss problem solving, thereby helping them to strengthen communication and critical-thinking skills, which are necessary for students to think analytically and reason. Mathematical language is being modeled by students' peers in their small-group discussions, and they will be sharing the various strategies they use to solve problems.

Models, Manipulatives, and Technological Tools

Working with materials in multiple representations strengthens students' understanding of the mathematics they are actively learning and provides a context for many abstract ideas. These tools, including calculators, help students think through a problematic situation; they provide a concrete or pictorial way to prove a conjecture; and they assist students to communicate their reasoning about a concept. Provide students with easy access to tools and expect them to assume responsibility of the materials they use daily. Technology can be used as a tool to enrich students' learning, and there are excellent problem-solving programs that encourage students' use of spatial reasoning. Several websites have been recommended for your students' use in the section titled "Resources to Support Teachers" near the end of this book. Although the use of models and manipulatives is critical in students' mathematics learning, it should not be the only means of instruction. Students also need to discuss what they are learning so that they will be able to connect their learning with abstract ideas about mathematics.

Reflecting on Problem Solving and Reasoning

As part of your classroom expectations, encourage all students to reflect on any mathematics they are learning, especially how they are thinking about a problem, the strategies they are using to solve it, and its solution. Students must learn to ask themselves, "Does this answer make sense?" This helps students self-monitor their thinking, and it allows them an opportunity to revise their reasoning if necessary. A supportive and collaborative classroom environment promotes reflective thinking.

High Standards and Expectations

All students have a right to be challenged and to make sense of the mathematics they are learning. "All students, regardless of their personal characteristics, backgrounds, or physical challenges, must have opportunities to study and support to learn mathematics" (NCTM 2000, 12). All students should participate in experiences in which they are reasoning about the mathematics they are learning.

Final Thoughts

As you reflect on this chapter, remember to set reasonable expectations when you begin to transition to a learning culture that promotes reasoning. The establishment of a positive climate is the most critical factor to consider in creating this environment, which is certainly a goal we want for all our students. We must clearly convey to students that we value their ways of reasoning and that we also expect them to respect and value each other's thinking and reasoning. We must also let students know that mathematics should make sense to them so that learning becomes meaningful.

Make it a priority to help your students understand that the process used to obtain a solution is just as important—if not more important—than the solution itself as a means to help them become powerful mathematical thinkers. And to share these processes, they will be expected to communicate their thinking and reasoning in a variety of ways. As teachers we often reflect on our own learning, and we must also expect our students to do the same. When we provide students time to think about their reasoning, refinements can be made that support them to communicate more clearly about their reasoning.

Another important goal to consider is ensuring the equity of educational opportunities for all our students. Students must be encouraged and supported as they begin to develop and use reasoning skills. Just as our students' understanding of concepts and skills are at different levels, so will be their reasoning abilities. Many of the "Classroom-Tested Tip" boxes throughout the chapters provide suggestions that will help you manage the diversity of your students' learning needs.

Although the transition to becoming a reasoning classroom may not be accomplished quickly, the journey is a worthwhile one. Provide your students multiple opportunities to reason about important mathematical ideas and relationships. These experiences help students develop flexible habits of mind. Helping our students to become powerful thinkers who reason thoughtfully and make sense of what they are learning will certainly prepare them for higher learning and their future roles in life.

Questions for Discussion

1. How do you personally feel about mathematics? How will this affect your students' learning of mathematics?

2. What issues and challenges have you experienced in reasoning about mathematics? What supports did you find helpful to enhance your ability to reason?

3. Think about how your students will feel about being actively involved in the learning of mathematics. What will you do to make this transition easier?

4. Which of the factors that help promote reasoning are considerations you are already implementing in your teaching of mathematics? How are they working? What changes do you want to make?

5. Will any of these factors be difficult for you to begin implementing into your daily teaching? Why do you think so? Talk with colleagues about strategies that will make the transition process more manageable for you.

Reasoning in Grades 6 Through 8

Mathematical reasoning is a part of mathematical thinking that involves forming generalizations and drawing valid conclusions about ideas and how they are related.

—Phares G. O'Daffer and Bruce A. Thornquist,
Research Ideas for the Classroom: High School Mathematics

What Can You Expect of Students?

Students in grades 6 through 8 should have had numerous opportunities in the elementary grades to make conjectures and reason about mathematical relationships as well as develop and justify simple mathematical arguments. They should have discovered that more than several examples are needed to prove a conjecture true, and that testing endless examples is not always necessary. As students progress through the middle grades, they should learn to justify true conjectures with deductive mathematical arguments and learn that a single counterexample can disprove an invalid conjecture. Making conjectures about mathematics concepts and skills becomes a regular occurrence, along with exploration of conjectures to determine whether they are valid or invalid.

Throughout the middle grades, students should be reasoning and generalizing about classes of objects or examples (e.g., conjecturing about properties of all rectangles or parallelograms, or reasoning about algebraic relationships). We can help students reason about classes of examples by asking questions such as the following:

- Will this work with other numbers?

- What do you think will happen if you solve ten more problems this way?

- Do you think this method will work every time?

Exploring a mathematical conjecture thoroughly provides students an opportunity to refine and edit it, and their reasoning skills are strengthened when they explore a conjecture with additional problems or numbers. These experiences help them develop more thoughtful generalizations and stronger justifications for their mathematical arguments than when they have no opportunity for such exploration.

Types of Reasoning: Inductive and Deductive

Middle grades students may still use intuitive reasoning in some situations, or reasoning based on feelings. However, throughout the middle grades we need to help students transition from intuitive reasoning to inductive and deductive reasoning. In inductive reasoning, students' thinking proceeds from specific ideas or examples to a general case or a pattern. Important opportunities for students to reason inductively are experiences in which they are able to recognize, extend, and generalize patterns as well as write rules using words or symbols to describe these patterns. In deductive reasoning, the student reaches a specific conclusion based on known information. Ideas and ways to implement inductive and deductive reasoning are provided in this chapter, including sample dialogues of students using both types of reasoning.

In Chapter 7, we discuss how students reason differently across the content standards. Examples of students using algebraic reasoning, spatial reasoning, and proportional reasoning are demonstrated in the student dialogues in Chapter 7. The student dialogues are discussions of activities included on the CD for each of the following content standards: number and operations, algebra, geometry, measurement, and data analysis and probability.

Inductive Reasoning

Patterns are the heart of mathematics, and inductive reasoning provides the vehicle for students to discover this fact. When students reason inductively, they explore important mathematical ideas about patterns, relationships, and generalities. Students benefit from our support in helping them explore statements or conjectures to determine the likelihood that they are always true by testing the conjecture with a variety of examples. However, we also want students to realize that testing with specific examples, no matter how many, is not enough to know that the conjecture is always true. Knowing something is always true requires deductive reasoning to generate a proof for all cases. In contrast, finding one counterexample to a conjecture is enough to disprove it. Too many students think one counterexample is just an anomaly to a statement that is otherwise true (Burke 1984; Galbraith 1981).

Students are likely to have had experiences working with patterns throughout the elementary years. In many of those experiences, students may have written informal rules, describing recursively how to obtain one term in a pattern from the previous term. They may have described the pattern only in words. Their rule may or may not be efficient at predicting future terms in the pattern. For instance, suppose a student describes the pattern 2, 5, 8, 11, 14, . . . as *add 3 to the previous term*. To find the 100th term in the pattern with this rule, students first need to find the 99th term. So, in the middle grades, students should be encouraged to use variables to describe the pattern with an explicit rule (in this case, $3n - 1$, where n is the number of the term in the sequence) because an explicit rule makes it efficient to find the 100th or 1000th or 1,000,000th term in the pattern simply by substituting the appropriate value for n. Explicit rules often make it easier to graph the pattern so that other important relationships within the pattern become evident. For instance, when *output* $= 3n - 1$ is graphed, students can recognize the pattern as linear with 3 as the rate of change from one term to the next.

Activities to help students develop inductive reasoning are easily implemented into a classroom instructional routine. Some of these may already be familiar to you. Textbooks often give rules for a pattern and expect students to determine the first few terms. Instead of giving students a rule, teachers can encourage inductive thinking by asking students to extend terms in a pattern and then generate a rule that describes the pattern. With this simple change, students have an opportunity to explore inductive thinking. If they are also encouraged to provide a general justification for their rule, inductive reasoning can extend to deductive reasoning as well.

Consider the *Build That Figure* task, shown in Figure 2–1, in which students are expected to extend a pattern and then provide a rule for the number of rectangles in the *n*th figure.

Teacher: Let's take a look at the pattern shown by these three figures. What do you notice?

Student: There's always 4 rectangles across the top.

Student: The number of rectangles down the sides is 1 more each time.

Teacher: Explain that a bit more. What do you mean when you say the number is "1 more each time"?

Student: The first figure has 3 down on each side, the second has 4 down on each side, and the third has 5 down on each side. So the number down on each side is 1 more each time than the previous one.

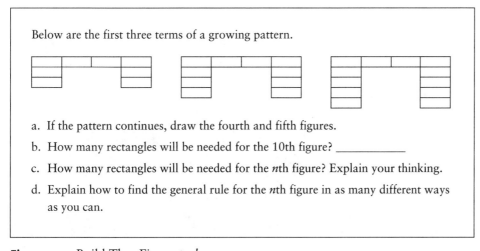

Below are the first three terms of a growing pattern.

a. If the pattern continues, draw the fourth and fifth figures.

b. How many rectangles will be needed for the 10th figure? _____

c. How many rectangles will be needed for the *n*th figure? Explain your thinking.

d. Explain how to find the general rule for the *n*th figure in as many different ways as you can.

Figure 2–1 Build That Figure *task.*

TEACHER: Do the rest of you agree with this thinking? (*Students nod in agreement.*) Let's have someone draw the fourth and fifth figures on the board while the rest of you draw the figure at your desk on your individual whiteboards. (See Figure 2–2.) Hold up your boards and let's see how many of you have the same figures. It looks like we agree with what's on the board.

TEACHER: Now suppose we want the number of rectangles in the 10th figure. Anybody have any ideas?

STUDENT: For Figure 10 there are 26 rectangles, since there are 4 across on the top and 12 down on each side. Since Figure 5 has 6 down, it would be doubled.

TEACHER: I am not quite sure I understand what you mean. Can you clarify your thinking?

STUDENT: Figure 5 has 6 down each side. So Figure 10 would have 12 down each side. 12 and 12 is 24 and then 4 across the top. That would be 28. Wait, that doesn't work.

TEACHER: Can anybody help us figure out what's wrong here?

STUDENT: You can't just double things because the figure numbers are doubled. But I can make his pattern work. On Figure 5, if you take out the 4 across the top, then what's left is 6 on each side. 6 is 1 more than the figure number 5. On Figure 4, if you take out the 4 across the top, then what's left is 5 on each side. So the number on each side is 1 more than the figure number. For the 10th figure, there would be 11 on each side and then 4 across the top. 11 + 11 + 4 is 26.

TEACHER: I think we have an idea here for a way to write a general rule. If we let n be the figure number, then how could we write the rule that was just described?

STUDENT: $(n + 1) + (n + 1) + 4$.

TEACHER: We have one way to write a rule for the pattern. Did anyone see the rule a different way?

Figure 2–2 *Student diagrams of the fourth and fifth figures in the pattern.*

STUDENT: I wrote $2n + 6$ because you start out with 6 blocks and 2 are added each time you go up a number. In the first figure, there are 4 blocks on the top and then 2 in the second row. Every figure has those. Then the first figure has 2 more rectangles. The second figure has 4 more rectangles, which is 2 times 2. The third figure has 6 more rectangles, which is 2 times 3. So the number of rectangles is 2 times the figure number plus 6, or $2n + 6$.

TEACHER: So we have a second way to write the rule. Is there another way to think about how the figure grows?

STUDENT: I imagined cutting the figure in half down the middle. Each side has the same amount. The number on each side is 3 more than the figure number. So, my rule is $2(n + 3)$.

TEACHER: We now have three ways to write the rule. Are there any other ways?

STUDENT: My way is like the first way but a little bit different. There are 2 rectangles in the middle of the first row. Then each side has 2 more than the figure number. So my rule is $2(n + 2) + 2$.

TEACHER: We have four different ways to write a rule. Are all of these ways equivalent? (See Figure 2–3.) (*The teacher now leads the class in a discussion of using the distributive property to show that all four representations are equivalent.*)

In this task, students were learning to reason inductively by extending a pattern and trying to generalize a rule by observing how the number of rectangles related to the figure number. Initially the students reasoned recursively when they described the number of rectangles on a side being 1 more than in the previous figure. Students also reasoned from a specific figure when they described how to find the number of rectangles in the 10th figure. Notice, however, that the teacher used students' descriptions to help them reason a general rule and describe it regardless of the figure number, thus moving from inductive reasoning to deductive reasoning. Also, the teacher validated a variety of ways to express the rule, helping students to make connections between their work with patterns and other work they had previously done with the distributive property. Such extended discussions are possible in classrooms in which the teacher has established a supportive environment where students are willing to try to clarify the thinking of their peers and provide alternative approaches.

$(n + 1) + (n + 1) + 4$

$2n + 6$

$2(n + 3)$

$2(n + 2) + 2$

Figure 2–3 *Different representations of the rule for the pattern.*

Lannin, Barker, and Townsend (2006) describe an extension to the type of reasoning illustrated in the student dialogue. Teachers might provide students with sample justifications to describe a rule for a pattern similar to that in the *Build That Figure* task. Students are then expected to evaluate each response to determine which responses provide a convincing argument that the rule is appropriate. Such variations not only help students strengthen their inductive reasoning skills but also build bridges to the development of deductive reasoning.

C L A S S R O O M - T E S T E D   T I P

Make it a priority to always ask "How do you know?" when students respond to a question, whether their answer is correct or incorrect. This gives students an opportunity to reflect on their response. If the answer is correct, others will benefit from the student's explanation of how he arrived at the answer, what strategy he used, and why the solution makes sense. If the answer is incorrect, the student has an opportunity to revise and edit any errors in his problem solving and reasoning, and other students will also have an opportunity to think about and analyze the errors. We often correct our students' errors before they can reflect on them. Students should see that errors are valuable learning experiences. It is essential that we allow our students multiple opportunities to analyze errors in problem solving and reasoning. This will help deepen their understandings of the mathematics they are learning.

Deductive Reasoning

Deductive reasoning is necessary for students when testing conjectures that have resulted from inductive reasoning. In the *Build That Figure* task just discussed, students engaged in deductive reasoning when they provided a general rule based on an overall description of how the pattern builds in terms of the figure number. That is, rather than focus on just the tenth figure, students connected the number of rectangles to any figure number—describing, for instance, the pattern as one more rectangle on each side than the figure number. They moved from a specific example to a generic example.

Puzzles or riddles provide another venue in which students can explore deductive reasoning and develop proofs, pictorially as well as symbolically. Consider the following *Mystery Number Riddle* task.

> **Choose a number between 1 and 10. Multiply it by 3. Add 2. Add the original number. Add 4. Divide by 2. Subtract 3. Record your original mystery number and your final result in the table at right.**

Students tried the riddle with several numbers. In each case, they applied the clues to the number they chose, told the teacher their final result, and had the teacher guess their original number. After the students' interest was piqued because the teacher

Original Mystery Number	Final Result after Clues
5	10
2	4
7	14
9	18
6	12

Figure 2–4 *Student table for* Mystery Number Riddle *task.*

read their mind and identified their original number, students recorded the mystery numbers and final result in a table. (See Figure 2–4.)

TEACHER: Let's review our work on the riddle. Look at the values in the table. How do the final results after the clues relate to the original numbers?

STUDENT: It doubled from the original number.

TEACHER: Do you think that will always happen? What if we tried a really small number or we tried a million?

STUDENT: Since it worked with the first five numbers, why wouldn't it work with the others?

TEACHER: That's a good question. Just because something has worked for five cases doesn't guarantee that it will work for every single case we could think of. The only way we can be sure that the riddle always doubles our original number is to write a general argument that works regardless of the number we choose. Sometimes we've used a letter like *n* to represent any number. Right now I'd like to use a box ☐ to represent our mystery number. So, based on our table (see Figure 2–4), what would the box be?

STUDENT: It would be 5, or 2, or 7, or 9, or 6.

TEACHER: That's right. But using a box for the number means it could be anything. When we add or subtract a number, we're going to use a dot • to represent what we add or subtract. I'd like a volunteer to come to the board and show us what the clues would look like when a box is our mystery number. (*Student comes to the board and illustrates the clues step by step as in Figure 2–5.*) So what does the last line of the diagram tell us?

STUDENT: Whenever you start with a box, you always get two boxes at the end. So the answer is always two times what you started with.

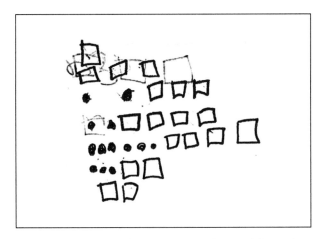

Figure 2-5 *Proof of* Mystery Number Riddle *task.*

In this case, the teacher used pictures to help sixth-grade students provide a visual deductive argument. Although these students had previously used variables to write rules for patterns, she decided to use a visual model for the proof rather than use variables to avoid having to address common errors that students make (e.g., incorrectly combining $n + 2$ as $2n$). She wanted to focus on reasoning without algebraic manipulation posing difficulties. As students progress into formal study of algebra, the same riddles can be used to provide reasoning opportunities within an algebraic context, and students can use n to represent any number.

The teacher in the previous dialogue provided initial clues to help students represent an argument deductively, as they had not previously had experience with this type of formal argument (e.g., using a box to represent the mystery number). Her clues helped provide a bridge from inductive to deductive reasoning. With more experiences of this kind, these students will be able to solve more complex problems independently and provide deductive arguments for their inductive reasoning about similar riddles. (See Figure 5–6, in which a student proves the same riddle using a variable for the mystery number.)

Another tool that helps students organize and visualize their thinking is a Venn diagram. A Venn diagram sorts the data from a problem so that students can make sense of it. Middle grades students may already be familiar with Venn diagrams from mathematics work in the elementary grades and from using these diagrams to compare the similarities and differences in stories in language arts.

CLASSROOM-TESTED TIP

If middle grades students have not previously worked with Venn diagrams, you may need to introduce students to simple diagrams consisting of a single circle or two non-overlapping circles before using two or three overlapping circles. In the dialogue that follows, students had previous experience with such figures and recognized that overlapping regions mean values belong to the groups represented by both circles.

In the following example, the teacher initially gave students the *Chicken, Beef, or Fish* task in Figure 2–6 and had them work collaboratively to gauge how they understood the problem. Then he brought the students together for a discussion of the problem.

TEACHER: (*At the overhead projector with a template of the Venn diagram and the student's numbers from Figure 2–6 on the diagram*) I have one solution on the overhead here. Based on this Venn diagram, how many people did the cafeteria staff survey?

STUDENT: 72, because that's what I get when I add up all the numbers.

TEACHER: Let's see if our diagram agrees with the student results from the survey. How many people liked chicken?

STUDENT: 19.

TEACHER: I see some other numbers inside the circle marked *chicken,* so it looks like more than 19 students like chicken.

STUDENT: The 19 are students who only like chicken because that part of the circle doesn't overlap with beef or fish.

TEACHER: Now I understand your thinking. But the problem doesn't say 19 only like chicken. So I think the 19 needs to include people who only like chicken as well as people who like chicken and also like beef or fish. How might we revise our solution so that the total of everything in the chicken circle is just 19 students?

The cafeteria staff at Marcella Middle School surveyed two classes of students about their choices of meats. Here are the results.

19 like chicken   4 like fish and beef
17 like beef   5 like chicken and fish
18 like fish   2 like all three
3 like chicken and beef   4 like none of the three

How many students participated in the cafeteria staff survey? Use the diagram below to show your work.

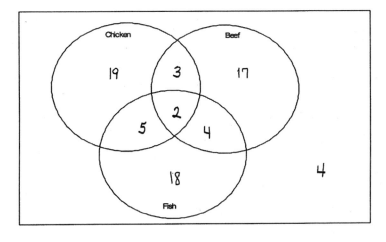

Figure 2–6 *Initial student work on* Chicken, Beef, or Fish *task.*

STUDENT: We could start with the part where all three circles overlap. There are 2 students who like chicken, beef, and fish. But I'm not sure what to do after that.

TEACHER: Can someone else help us out?

STUDENT: There are two parts that are in both the chicken and beef circles and we know 3 students like chicken and beef. We already put 2 in the part where all three overlap. So, 3 – 2 leaves 1 for the part that is chicken and beef.

TEACHER: What do the rest of you think about his reasoning? (*Students seem to be in agreement.*) What if we look at the parts that are chicken and fish?

STUDENT: We know that 5 like fish and chicken. Since 2 are in the part where all three circles overlap, that leaves 5 – 2 or 3 for the circle that is just chicken and fish.

STUDENT: Altogether we want 19 in the chicken circle. In the little parts we've got 3 + 2 + 1 or 6 so that means 19 – 6 or 13 just like chicken.

TEACHER: Let's see if you can revise your answers for the rest of the Venn diagram in your groups. (*Students then worked together to modify their responses for beef and fish and found a total of 48 students surveyed.*)

The teacher was not surprised that students initially read the 19 as applying to students who only liked chicken. In fact, the teacher chose this problem to help students learn to read carefully and not to infer words that were not there. As students reason, they need to be careful that they do not make unsubstantiated assumptions.

Venn diagrams have several connections to other topics in the middle grades curriculum. Consider a situation with two groups *A* and *B*. The number of elements in the union of *A* and *B* is the sum of the number of elements in each of *A* and *B* less the number of elements in the intersection of *A* and *B*. Counting problems arise not only in situations like those in the previous student dialogue but also when studying probability. Suppose a jar contains the numbers 1, 2, 3, 4, 5, 6, 7, 8, 9, and 10. One number is to be chosen at random from the jar. What is the probability that the number chosen is either odd or a prime number? There are 6 possibilities that satisfy the conditions (1, 2, 3, 5, 7, and 9), so the probability is $\frac{6}{10}$. The probability could be found by adding the probability of selecting an odd number ($\frac{5}{10}$) to the probability of selecting a prime number ($\frac{4}{10}$) and then subtracting the probability of selecting a number that is both odd and a prime ($\frac{3}{10}$). Although the given problem could certainly be solved without applying the use of Venn diagrams or unions, there are times when the underlying concepts of Venn diagrams can help students to reason about the situation.

Consider the following *Skateboarding or Bicycling* task.

The 23 students in Richard's class are asked the following question: "Do you like skateboarding or bicycling?" Of these, 15 students say they like bicycling and 11 say they like skateboarding. How many like both bicycling and skateboarding?

TEACHER: How might we use a Venn diagram to help us answer this problem?

STUDENT: If I add the 15 who like bicycling and the 11 who like skateboarding, I get 26 which is too many. This must mean there are 3 who like both.

TEACHER: Why can you say that?

STUDENT: Well, when you add all of the skateboarding and all the bicycling, the part where they overlap is counted twice. So, if I put 3 in the intersection, then there are 15 – 3 or 12 who only like bicycling and 11 – 3 or 8 who only like skateboarding. When I add 8 + 3 + 12, I get 23 so I know my answer is correct.

In this dialogue, notice that the use of a visual model through a Venn diagram helped students reason about the overlap in counting. We should never discount the power of visual models to help students reason mathematically.

Providing students multiple opportunities to solve logical-thinking problems helps them to develop the thinking that is foundational for more complicated types of reasoning in later school years. In this chapter, we have discussed several ways to do this. Using number logic problems, building patterns, riddles, or counting problems is one way to support students in building their inductive and deductive reasoning abilities. These logic problems are generally easy to generate, and once students have become comfortable in solving the problems you develop, you can ask them to create their own logic problems to present to classmates to solve.

In the process of discussing and solving problems similar to the ones in this chapter, students

■ strengthen their number sense;

■ work on a challenging problem; and

■ use deductive reasoning skills.

Additional Tools That Support Students' Reasoning Skills

Teachers find that games are highly motivating for students. Mathematics games help students develop number sense and encourage them to think about and then use multiple strategies as they are playing. Even when our students are playing games, we can learn important information about their reasoning abilities. Give yourself time to observe and ask questions of your students as they play math games with classmates. In addition, the discussions you have with students following any game playing are critical and beneficial for your students as they develop the skills necessary for reasoning.

One example of a logical mathematics game that many find interesting and challenging is a Sudoku puzzle. In addition to books containing numerous Sudoku puzzles, many newspapers run one puzzle a day, with increasing complexity over the course of the week. Sudoku puzzles contain nine sections to form a 9×9 grid overall, with each section containing a 3×3 square. In each 3×3 square, each of the numbers 1 through 9 must be placed. However, in each of the 9 rows and each of the 9 columns the numbers 1 through 9 can only be placed one time. Each puzzle is formatted with a different number of cells completed or empty. Both inductive and deductive reasoning must be used to complete the missing cells.

Final Thoughts

Student discussions are essential in any mathematics lesson you present to your students. Simply asking students to complete a problem-solving activity without giving them an opportunity to discuss their problem solving and reasoning deprives them and us of a valuable learning experience!

Encourage your students to engage in the type of discussions in which they, rather than you, are held accountable for talk. This allows students to see that their thinking is the focus of the discussion. As students begin to listen thoughtfully to each other, they will learn to ask questions of their classmates to help them better understand the reasoning that is being shared. Students will be encouraged to justify answers using examples from their problem solving when they know their contributions are valued. Students in a classroom climate that is safe, respectful, and encourages learning in a collaborative manner are more willing to challenge others' reasoning, which aids in the development of reasoning skills.

Inductive and deductive reasoning are important aspects of reasoning. It is through the discussions we have with our students, as well as our expectations for students to write and represent their reasoning, that we can support them in developing their ability to reason. We must expect our students to explain their reasoning in thinking about a task, whether the solution is incorrect or correct. Asking, "How do you know?" is a valuable instructional move! Students should understand that incorrect responses are also important to their learning and are beneficial in helping them refine their reasoning.

The varied questions modeled by teachers in the student dialogues of this chapter are examples of how to initiate more reflective discussion from your students. The questions helped students reason using inductive and deductive thinking skills. Remembering a list of questions may be overwhelming to you when you begin to have daily discussions with your students. Choose a few questions to ask in your initial discussions. This will help you become comfortable with the questions, and students will hear the same questions consistently modeled. Knowing that the questions will be asked routinely helps them become comfortable with math talk. Once you are familiar with these initial questions, add additional ones to your repertoire. Not only are questions modeled for you throughout this book's dialogues, but they are also listed in some of the "Classroom-Tested Tip" boxes in Chapter 1. Additional articles listed in the Teacher Resources near the end of this book provide further examples of teachers questioning students as they reason about rich mathematics tasks.

Questions for Discussion

1. How are the types of reasoning discussed in this chapter different? Which of these types of reasoning did you have the most opportunities to use during your own education in middle school? How did that affect your mathematics abilities in later school years?

2. What challenges do you expect to face as you support your students in using the types of reasoning discussed in this chapter?

3. Think about, or discuss with your colleagues, how you might enhance or begin to change your instruction to include more reasoning opportunities for your students. What experiences will you provide your students to help them strengthen their reasoning abilities?

4. Review your instructional materials. What types of reasoning opportunities are included in them? How frequently do they appear? In what ways will you need to supplement your instructional materials to provide opportunities for your students to reason?

Is It Always True? Making Conjectures

*A conjecture is a guess, inference, theory, or prediction based on untested
or unproven—and hence, uncertain or incomplete—evidence.*

—Arthur J. Baroody and Ronald T. Coslick,
Fostering Children's Mathematical Power

The Role of Conjectures in Reasoning and Proof

Students in the middle grades are capable of making, editing, refining, and testing a
variety of mathematical conjectures. Their experiences with patterning in the elementary grades were foundational for this type of reasoning. In addition, the thinking necessary for making conjectures was being developed when students were asked to
predict what would happen next in a problem-solving situation and why it would happen. When students make conjectures about mathematical relationships, they examine these relationships more thoroughly. This in turn enables students to gain a deep
understanding of the mathematics being learned.

Making conjectures is a critical component of reasoning. Make it a habit to ask
students when appropriate, "Is this always true?" Asking this question helps students
learn to develop and justify conjectures that are based on broad generalizations. When
students make conjectures, they need to know that these conjectures should be supported by informal mathematical arguments or justifications. To validate conjectures,
students must use deductive reasoning to develop their mathematical arguments. But
before students attempt a deductive argument, they need to explore the potential truth
of a conjecture by examining many examples. If they find one example that does not
satisfy the conditions in the conjecture, they have disproved it with a counterexample. If they find many examples for which the conjecture is true, they have not yet proved

it is true but can be encouraged to use deductive reasoning to provide an appropriate argument. Provide a variety of materials and tools to aid students when they investigate conjectures, such as manipulatives, calculators, and measuring tools (rulers, balances, etc.). It is important to encourage students to represent their mathematical arguments pictorially and/or symbolically. Because students in grades 6 through 8 are at different levels of fluency in their written communication skills, whether or not they are native English speakers, it is essential that teachers accept arguments in a variety of formats.

As students begin to reason more about the mathematics they are learning in your daily discussions, they will encounter reasoning that doesn't make sense or is flawed. Our natural reaction is to correct a student's flawed reasoning and continue with the lesson, but we actually deny students a valuable learning opportunity when we do this. In a classroom that promotes reasoning, students' misconceptions and flawed reasoning are critical pieces of information about how our students are thinking about a mathematical idea or relationship. We want our students to understand that flaws in reasoning should be viewed as a natural part of the reasoning process.

At first it may be difficult for students to identify any flaws in their own reasoning; rather, they often find it easier to explain flaws in someone else's reasoning. Communicate to students that their reasoning may be challenged, and that when this occurs, these misunderstandings and flaws are a necessary part of reasoning. They should also understand that flawed reasoning will be examined with the intent of learning from it. This experience provides students an opportunity to reflect, analyze, and rethink the proposed conjecture, and helps students know what it is like to think through reasoning that does not work or make sense. The probing questions you ask will support students in reexamining reasoning that is being shared, and students should be encouraged to revise any flawed reasoning that is identified during a discussion. Students who are comfortable with rethinking about reasoning will be prepared to self-monitor their own thinking when similar problem-solving situations arise in later school years.

In the middle grades, students' prior belief that something is true simply because it happened before begins to shift toward a realization that conjectures must be justified in order to prove their validity. Thinking about reasoning in this way and questioning, testing, and revising conjectures are experiences all our students deserve. Yet, such experiences are often lacking in textbooks. So, we must provide students with multiple opportunities to explore the validity of claims. In this chapter, we examine how to provide students with experiences that encourage reasoning about and making conjectures.

Many mathematical relationships that students have studied from a numerical perspective in the elementary grades, such as the commutative or distributive properties, are now candidates for deeper study from a symbolic perspective. In this chapter, we illustrate how students can be encouraged to reason about properties they have used throughout the elementary grades. We provide questions to ask your students that will assist them in thinking about their reasoning. These questions focus students' thinking, which helps them communicate conjectures in a way that makes sense and clearly conveys the meaning of the generalizations. As in previous chapters, examples of student dialogues are included to model how this type of reasoning could look in your classroom. "Students must practice looking for relationships, making conjectures, testing their conjectures, and explaining and justifying the generalizations they make" (Russell 1999, 10).

How Can You Help Your Students Reason About Conjectures?

In order for students to begin the process of making conjectures, it is critical that an accepting classroom environment has been well established. We want our students to view the process of making and sharing conjectures as a risk-free and comfortable experience. When you are beginning to support students in making and stating conjectures, we recommend that you first revisit Chapter 1. The factors described in that chapter are those that promote a reasoning classroom and will certainly make this transition smoother for you and your students.

Students articulate conjectures in words when they first learn how to conjecture. Over time with multiple experiences, they find that symbols more precisely express their conjectures. Students who have experienced limited opportunities to reason and explain reasoning may find it difficult to state a conjecture about a relationship they observe. Our support becomes critical in guiding students' thinking about mathematical ideas, patterns, and relationships they reason about, so that their conjectures become more precise and understandable. This ability evolves over time with repeated experiences.

The questions we ask of our students assist them in learning how to articulate conjectures, and these questions guide them to consider how to state conjectures that are clearly understood. Students must know that it is acceptable to have a conjecture questioned, clarified, revised, and tested. This sends an important message to our students that their conjectures will be carefully examined and revised if necessary. Understanding this expectation helps students become more comfortable when sharing a conjecture with their peers.

A strategy for introducing students to the reasoning involved in making a conjecture about a generalization is to show students a mathematical sentence and simply ask, "Is it true?" In answering this question, students must examine the sentence or equation closely. They might start by checking one or more examples. A simple equation or property can provide an opportunity for reasoning in the middle grades, particularly if the property is stated symbolically. Consider the *Is It True When Dividing* task that focuses on whether or not division is commutative:

> **For all real numbers, r and s, with $r \neq 0$ and $s \neq 0$, is it true that $r \div s = s \div r$?**
> **_____ Yes _____ No**
> **Suppose you had to convince a person in another class that your answer is correct. Explain your reasoning.**

TEACHER: I want you to take a minute and look at the sentence I have on the overhead. (*The teacher has a copy of the task displayed for students to see.*) What do you think about this statement? What is the sentence asking?

STUDENT: If you can switch the order when you divide.

TEACHER: Does anyone remember a name for the property that lets us switch the order of a computation? (*Teacher waits a few seconds, but there is no response.*) Remember when we talked before about the commutative property when we add or multiply? (*Students nod.*) This sentence gives us a chance to decide if we think division is commutative as well. Is the sentence ever true?

STUDENT: Do r and s have to be different numbers?

TEACHER: Good question. What do you think?

STUDENT: The problem doesn't say they have to be different. If r is 3 and s is 3, then r divided by s is 3 divided by 3 which is 1. You get the same thing when you do s divided by r. So the sentence is true.

STUDENT: Anytime r and s have the same value, the sentence is true because both sides equal 1.

TEACHER: That's a great observation. So, you've suggested the statement is true because we can find lots of examples where both sides of the equation give the same value. We just need to choose r and s to be equal to each other. Have we met all the conditions of the problem?

STUDENT: The question asks if it's true for all values of r and s that aren't 0. So I think you have to try some cases when r and s aren't the same. When I tried $r = 2$ and $s = 4$, it didn't work.

TEACHER: Come show us what you did. (*The student brings his paper and displays it using a document camera, or Elmo, so that the class can see his work. See Figure 3–1.*) So, he's given us an example where the statement doesn't work. Anybody else have an example where the statement doesn't work?

STUDENT: I tried $r = 9$ and $s = 3$. (See Figure 3–2.)

TEACHER: We now have two examples that show the statement is not true. So, what can we conclude about this statement?

STUDENT: It's not true. Even though we found some cases when it works, it doesn't work for all cases, so it's not always true.

TEACHER: Let me ask you, did we really need both examples (Figures 3–1 and 3–2) to disprove the statement?

STUDENT: We only needed one case. But the second case helps us make sure we didn't make a mistake when we did the first example.

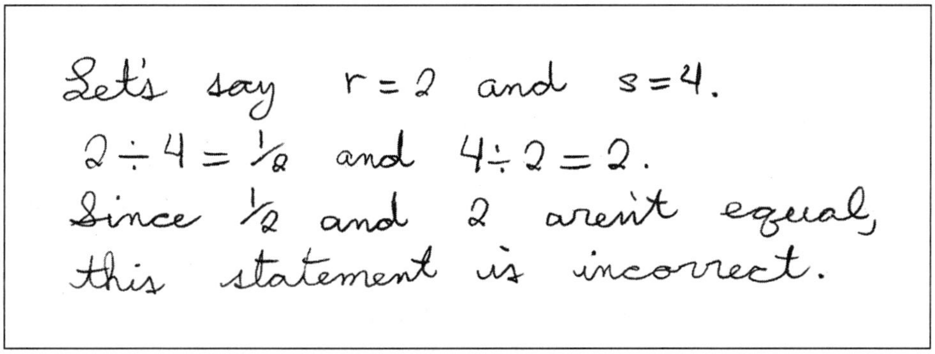

Let's say $r = 2$ and $s = 4$.
$2 \div 4 = \frac{1}{2}$ and $4 \div 2 = 2$.
Since $\frac{1}{2}$ and 2 aren't equal, this statement is incorrect.

Figure 3-1 *Student response to task about whether division is commutative.*

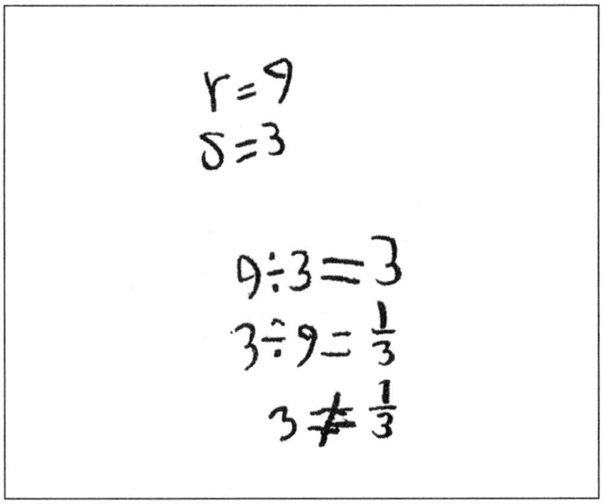

$$r = 9$$
$$s = 3$$

$$9 \div 3 = 3$$
$$3 \div 9 = \frac{1}{3}$$
$$3 \neq \frac{1}{3}$$

Figure 3–2 *Second student response to task about whether division is commutative.*

TEACHER: Because the statement had to be true for all values of *r* and *s*, you're absolutely right that finding just one example that doesn't work is enough. By the way, why do you think 0 was not allowed for *r* or *s*?

STUDENT: Because you can't divide by 0.

Tasks like the one in the dialogue are important in helping students think about important algebraic properties in a general manner, in this case the commutative property. Division is particularly problematic because students often verbalize division expressions incorrectly. For instance, it is not uncommon to hear students read $4\overline{)12}$ as "4 divided by 12," even when they interpret the result correctly as 3. The division format that students see frequently in the elementary grades, $4\overline{)12}$, may contribute to this difficulty with verbalization because the order of the 4 and 12 when using the "divided into" symbol must be reversed when rewritten using the "divided by" symbol (÷) (12 ÷ 4). Paying attention to the order of the division is important when using a calculator to do a computation, such as $12\overline{)288}$. Students often want to enter the computation in the calculator as they see it (12 ÷ 288) rather than as it needs to be interpreted (288 ÷ 12).

The teacher in the dialogue drew attention to several issues that often arise when students explore conjectures. First, the teacher had students consider any cases in which the statement might be true. There are many times when a conjecture may be true for a few cases but not true in general. Too often students use special cases, such as 1 or the same values for both variables, rather than a general case. Drawing students' attention to special cases and then helping them realize limitations of that approach is important to strengthen their reasoning abilities.

Second, the teacher validated that there would be multiple examples that could disprove the conjecture by having different students demonstrate their counterexample to the statement. However, the teacher drew attention to the fact that only one counterexample is needed. Sometimes students feel more comfortable finding multiple counterexamples, but they should realize that only one is really needed to disprove the statement.

Third, the teacher focused students' attention on restrictions for which the conjecture might be valid or not, namely that neither r nor s could equal zero. Conjectures often have constraints that limit the conditions under which they are true, and students need to attend to any such constraints.

Discussions like this do not occur if students are simply asked, "What is twenty-five divided by three?" This type of question requires only a one-word answer; instead the teacher asked the question, "Is this number sentence true or false?" The question produced a rich discussion about the commutative property. Students already had experience with the commutative property but now had an opportunity to extend their thinking to a new operation and to realize that properties are often true only under specific conditions.

CLASSROOM-TESTED TIP

Give students an opportunity to record their thinking and share with the rest of the class. If you have a document camera (often called an Elmo) in your room, then it is relatively easy to have students project the work from their paper onto a screen or a board for their peers to see. If you don't have such a device, you might consider having students write on transparencies and then display their work on the overhead. As students share their thinking and reasoning, provide them with opportunities to revise their work and to incorporate new knowledge and thinking into any written explanations.

True-or-false sentences are a useful tool to have students reason about important properties in mathematics. They are particularly beneficial to focus attention on common misconceptions that students have or errors that students make. A friend of one of the authors regularly says, "If you can't fix it, feature it!" As teachers, we often say over and over that certain properties are not true [e.g., $(x + y)^2 \neq x^2 + y^2$ or $\sqrt{x^2 + y^2} \neq x + y$ or $\frac{30 + x}{5} \neq 6 + x$ or $2x + 3x \neq 5x^2$], but students persist in making these errors. Common mistakes become candidates for true-or-false sentences and give us insight into students' thinking that can help focus our instruction. Several tasks of this nature are included on the CD.

Consider the *Is It True When Multiplying* task, which addresses a common error related to the distributive property of multiplication over subtraction.

For all real numbers 3, x, and y, is it true that $3(x - y) = 3x - y$?
_____ **Yes** _____ **No**
Suppose you had to convince a person in another class that your answer is correct. Explain your reasoning.

Figures 3–3 and 3–4 contain two sample responses to this task from students. Although the student response in Figure 3–3 suggests that the student recognizes the

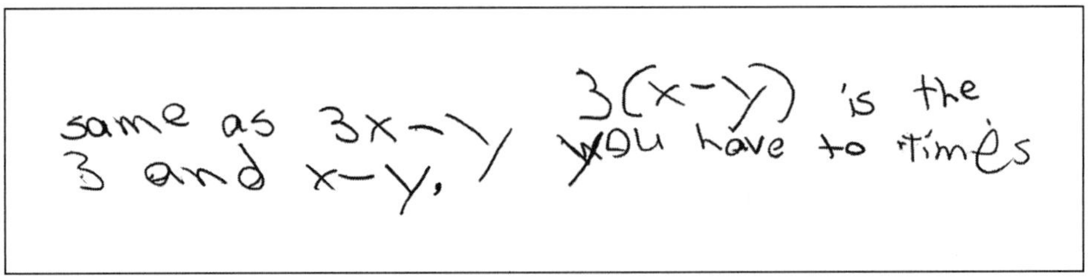

same as $3x - y$,
3 and $x - y$,

$3(x-y)$ is the
you have to times

Figure 3–3 *Student misconception about $3(x - y) = 3x - y$.*

meaning of the notation and that the 3 needs to be multiplied by $x - y$, the student is not able to carry out the multiplication correctly. Also, the student fails to try the sentence with any specific values for x and y. In contrast, the student response in Figure 3–4 demonstrates understanding about how to disprove a statement. Additionally, this student recognizes that certain numbers, such as 0, are often not good candidates as examples. Although it is wise to avoid 0 for either variable, in this case if $x = 0$ the statement is false even though it is true when $y = 0$. The student clearly writes an argument to show that the statement is not always true by evaluating each side of the equation for the given choices of x and y and then finalizes the argument by writing $-18 \neq 0$. This student is well on the way toward writing good deductive arguments. In contrast, the first student may still need help in determining how to proceed to assess the validity of a statement, namely by trying a few examples. Despite the misconceptions displayed by the first student, the teacher has insights into the student's thinking that would likely not have occurred if the problem had just been to evaluate $3(6 - 2)$.

Reasoning about important mathematical relationships need not always occur through abstract or symbolic statements such as the two examples discussed thus far. Sometimes, contexts provide an important entry for students to reason about relationships. Consider the *Find the Sale Price* task which provides a contextual example related to use of the distributive property.

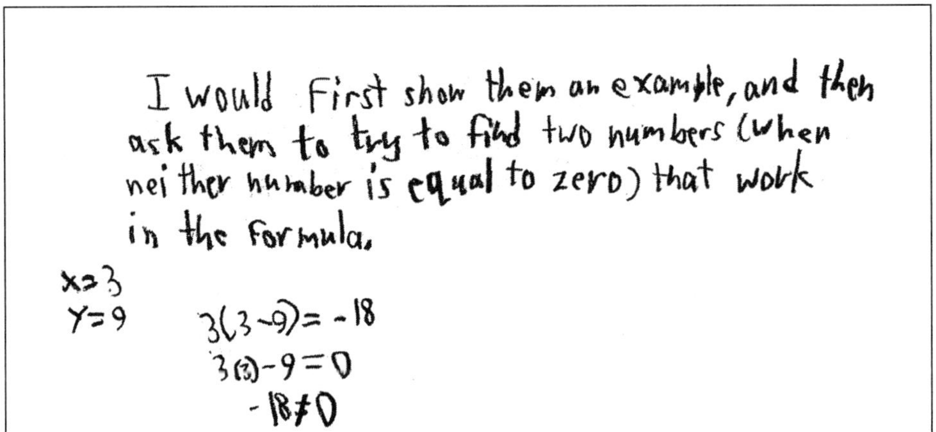

I would first show them an example, and then ask them to try to find two numbers (when neither number is equal to zero) that work in the formula.

$x = 3$
$Y = 9$
$3(3 - 9) = -18$
$3(3) - 9 = 0$
$-18 \neq 0$

Figure 3–4 *Student response to task about $3(x - y) = 3x - y$.*

A coat that usually costs $150 is on sale for 30% off the cost. How much will the coat cost on sale?

Sally and Rupert solved this problem in two different ways:

Sally: I multiplied $150 by 30% to get the amount taken off the price. Then I subtracted that amount from $150. That gives me the sale price.

Rupert: I multiplied $150 by 70%, and that gives me the sale price.

Who is correct? Explain your thinking.

TEACHER: I'd like you to consider the problem on the board. (*Teacher has the problem displayed on the overhead.*) Sally and Rupert have two different methods to find the sale price. Do you think both of them could be correct?

STUDENT: Maybe. Sometimes you can do a problem in a different way.

TEACHER: Take a few minutes and work together on the problem. Be prepared to explain who's correct and how you know. (*Teacher gives students a few minutes to work. Students have calculators available so that computational proficiency does not get in the way of thinking and reasoning about the problem.*)

STUDENT: We think they're both correct. (*Student shows the work in Figure 3–5.*)

TEACHER: What do the rest of you think about this group's work? (*Most students agree and had done similar computations.*) So, how would you use Rupert's method if the coat had been on sale for 40% off the original price instead of 30%?

STUDENT: Rupert will multiply the original price by 60% since he uses the amount that is 100% if added.

TEACHER: I need you to explain that a bit more. What do you mean by "the amount that is 100% if added"?

STUDENT: The amount taken off the price and the amount he multiplies by to find the cost need to add to 100% because 100% is the total. So, if the sale price is 40% off, then you subtract 100% – 40% and you have 60% left to pay.

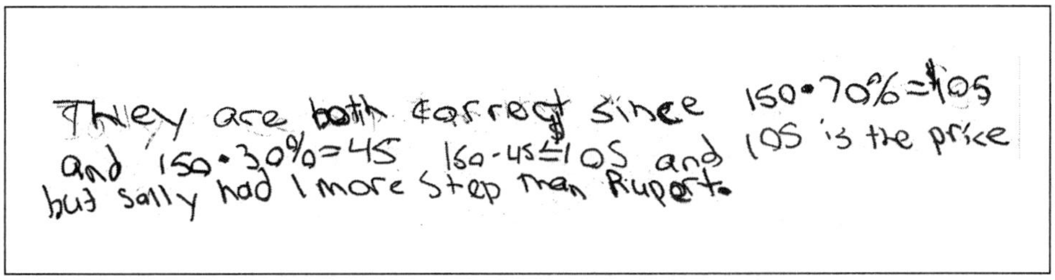

They are both correct since 150•70%=$105 and 150•30%=45 150-45=105 and 105 is the price but Sally had 1 more step than Rupert.

Figure 3–5 *Student work to compare Sally's and Rupert's methods for finding cost.*

TEACHER: Let's make a conjecture that explains Rupert's method no matter what the percent off the price. Who can state a general statement for what we just did in a specific case?

STUDENT: Subtract the percent off minus 100 and then multiply that by the original price.

TEACHER: How many people agree? (*The class is about evenly divided about agreement.*)

STUDENT: I think what he said is backwards. I think you should do 100 minus the percent off and then multiply by the original price. (*Students seem to agree with this revision of the conjecture.*)

TEACHER: (*addressed to the first student who stated the initial conjecture*) Are you okay with that revision of your statement? (*Student nods in agreement.*) Let's see if we can show Sally's method with symbols. Suppose you didn't know the original price. Let's call that price n. How could you show Sally's method? (*Teacher waits a few minutes for students to think about the problem.*)

STUDENT: You would multiply 30% times n and then subtract from n. (*Teacher writes $n - 0.3n$ on the board.*)

TEACHER: And what happens when we simplify that? (*Students seem puzzled.*) Remember that n can be written as $1n$, so we can write $n - 0.3n$ as $1n - 0.3n$. How could you use the distributive property to simplify this?

STUDENT: Both have n so you could put n outside the parentheses. That gives $n(1 - 0.3)$ or $0.7n$. That's just what we said earlier about subtracting from 100%.

STUDENT: That's just what Rupert did.

TEACHER: Exactly. You just used the distributive property to show why Rupert's method always works. Suppose Sally found a matching hat that costs $12. How could Sally use Rupert's method to find the total cost of the hat with a 7% sales tax?

STUDENT: With sales tax you have to add. So, I think you would have 100% + 7%, which is 107%. I just have to multiply the 12 by 1.07.

STUDENT: Rupert's method is shorter and easier because it's only one step, not two. I like Rupert's method better.

In this instance, students understood the problem and could reason about how to determine the percent needed to use Rupert's method no matter what the percent of discount. The teacher chose to help students provide a symbolic argument to justify why 30% off is equivalent to multiplying the original price by 70%. She was helping students bridge a verbal description of a conjecture to a more symbolic statement and to make connections to the distributive property they had previously studied.

Making Conjectures About Important Mathematical Ideas and Relationships

When you choose mathematics concepts or skills as contexts for your students to explore conjectures, it is critical to consider your students' mathematical strengths, needs, and understandings. In the previous section, we examined how the use of true-or-false sentences can help facilitate students' reasoning about conjectures. These equations are an excellent way to enable students to reason about algebraic properties and relationships.

In addition, it is valuable for students to generalize and make conjectures about the outcome of their calculations. The discussion of Rupert's method to find the sale price falls in this category, even though students were exploring the result of a calculation invented by a fictitious student. Creating conjectures that are eventually represented by symbols, as was done in the dialogue, is a goal to set for our students. It lays out important groundwork for later algebraic thinking.

In addition to conjectures about numerical or algebraic situations, opportunities for conjectures and reasoning occur throughout the study of geometry. Consider the questions and student responses in Figure 3–6 from the task *Obtuse Angles and Polygons* found on the CD. The wording of the questions is not in the typical form of a conjecture, but the questions ask students to determine the possibility of a particular property. (Contrast the statement *Can a triangle have more than one obtuse angle?* with a more explicit statement of a conjecture, *A triangle cannot have more than one obtuse angle.*) The questions encourage students to reason about properties of polygons and to draw a figure to illustrate the property if it is true or to explain why

1. **Can a triangle have more than one obtuse angle? How do you know?**

   No, because the three angles would add up to more then 180°.

2. a. **Can a quadrilateral have two obtuse angles? If so, draw an example. If not, explain why not.**

   obtuse

   b. **Can a quadrilateral have three obtuse angles? If so, draw an example. If not, explain why not.**

   c. **Can a quadrilateral have four obtuse angles? If so, draw an example. If not, explain why not.**

   no

Figure 3–6 *Student responses to* Obtuse Angles and Polygons.

the property is not true. In the first question, the student reasoned that more than one obtuse angle would yield a sum for the angle measures that is greater than 180°. For each of the first two parts of the second question, the student generated an example to illustrate why the question is true. In the third part of question 2, the student reasoned that the situation is not possible but failed to write an explanation. In discussion with the student, she recognized that the sum of the angle measures in a quadrilateral is 360°, so that the sum of the measures of four obtuse angles would be greater than 360°, and thus, not possible. In an extension to the task, the student was able to reason about the measures of angles in polygons with more than four sides to determine that an n-gon, with $n > 4$, can have n obtuse angles.

What Questions Help Students Think About Conjectures?

Specific questions can focus students in thinking and reasoning about a mathematical relationship, which then guides them in thinking clearly about a conjecture. Make it a priority to ask probing questions that will help your students reflect on their reasoning, especially when students' thinking needs clarification. Being comfortable in asking questions of our students is crucial. When we are consistent in the questions we ask, we provide a model that students can emulate when they ask questions about other students' reasoning and proposed conjectures. The following questions can guide students in thinking about generalizations and making conjectures about them:

- What can you say about this mathematical sentence?

- What is a conjecture you can make about this?

- How can we state a conjecture for this mathematical sentence?

- Is it always true?

- How do you know?

- Why does this work? Will it work every time?

- If we try this with more problems, will it still work?

- Does this conjecture make sense to all of you?

- How could we revise this conjecture so that it is stated more clearly?

- What would happen if you . . . ?

Here are examples of probing questions or prompts that further help to develop and clarify a conjecture:

- Why do you agree?

- Why do you disagree?

- What did you find out when you . . . ?

- Give us another example of why that works.

- I would like someone to explain to me what was just said.

- I need someone to restate that reasoning in a different way.

- I would like someone else to restate the conjecture using their own words.

- What do you mean when you say it is always true?

- Does this conjecture make sense to you? Why or why not?

- What do you think about this?

- What were you thinking about when you said . . . ?

Record conjectures that students make on the board or on transparencies and display them for others to see. After a conjecture has been agreed upon by students and recorded, reread it to students so they can evaluate the conjecture as stated. A goal for students is to make conjecture statements clear and precise. Conjectures that are not clearly stated will initiate counterexamples. Ask students questions that model how to self-monitor their own reasoning and how to think about the process of formulating conjectures. For example:

- Does the conjecture make sense to you?

- If someone was visiting our classroom and read your conjecture, would she understand what it meant?

- What could be added to this conjecture so that it would be clearer?

- What parts of the conjecture, if any, need to be deleted because they are confusing?

- Is there enough information in the conjecture so that it clearly explains the reasoning of the conjecture?

Students will learn that a conjecture must be precise and clearly stated so that no confusion occurs about its intent. If a conjecture is too general, it causes confusion, and counterexamples can disprove it. The ability to make specific, detailed conjectures happens over time with multiple experiences. We want our students to understand that it is natural to edit and revise conjectures in order to make them clearly understandable by all.

C L A S S R O O M - T E S T E D   T I P

Use the word *conjecture* with students. Explain that a conjecture is a guess, prediction, or idea based on unproven evidence. As you begin implementing the expectation that students make conjectures about mathematical ideas, relationships, and operations, encourage them to edit and revise them in order to make the conjectures more precise. This self-monitoring of thinking is an important skill for students.

Final Thoughts

When students make conjectures, they draw on their informal knowledge about mathematical ideas and relationships and articulate their thinking so that it becomes more explicit. With multiple experiences, students also learn that precise language is necessary to state a conjecture clearly to avoid confusion. Clarity happens when students are encouraged to reflect on their conjectures to refine them.

When students are encouraged to reflect on their thinking, they have the support needed to reason about generalizations, which helps to deepen their understandings of mathematics. Reflective thinking helps students learn to evaluate the reasonableness of the conjectures they make as they begin to develop mathematical arguments to test their conjectures. It is important for students to experience situations in which counterexamples show that a conjecture is incorrect and therefore a revision of a stated conjecture is needed.

As you plan instruction that enables your students to reason about mathematics, you will want students to make conjectures about important mathematical ideas and relationships. Choose ideas and relationships that provide your students with experiences that will help build the foundational understandings necessary for learning mathematics in later grades. Students' examination of the properties of operations, relations, and equality helps them understand why computation procedures work the way they do and enables students to connect arithmetic to algebra. These connections help them learn algebra with understanding.

The strategy of simply asking students whether an equation is true or false, and then why they think so, will lead students to make conjectures about properties of operations. Likewise, asking students about a property in geometry or a property of a shape provides a non-computational situation in which to reason. We need to provide students with opportunities to reason in a range of situations so that students realize reasoning is a natural occurrence throughout mathematics.

Questioning can focus our students so that they are able to identify generalizations and make conjectures. Although lower-level questions have a place in our instruction in some instances, it is the higher-level questions that initiate the kind of thinking our students use to make sense of what they are learning. These questions motivate students to think, reflect, analyze, or synthesize information. A question such as "Why do you think that is true?" motivates conjecture making.

The importance of a collaborative environment for developing students' reasoning skills should not be overlooked, and questions such as, "What does the rest of your group think about Nicholas' conjecture?" and "How will you convince us that your reasoning is valid?", encourage students to work together to make sense of mathematics. And a question like "How does this idea relate to what we learned in our measurement unit?" helps students connect mathematics and its applications.

Students initially use words to state their conjectures, as demonstrated in the student dialogue examples. In some of these dialogue examples, students clarified their word choices so their conjectures made more sense and were stated more precisely. As students' mathematical reasoning becomes more sophisticated, they see the need for describing conjectures using symbols to represent their mathematical reasoning.

Questions for Discussion

1. What kinds of experiences have your students already had with making conjectures? How will you build on these experiences to enhance your students' reasoning skills?

2. True-or-false sentences were introduced in this chapter as a way to support students in making conjectures. What are several true-or-false sentences you could present to your students to begin this process? If possible, discuss these with your colleagues.

3. The importance of students making conjectures about relationships was also discussed in this chapter. On which number properties or relationships would you like your students to focus in order to make generalizations, or conjectures, about them?

4. How comfortable are you with the suggested questions in this chapter? Which questions would be the easiest to implement for you and your students? How do you see your students' role in student discussions changing in regard to questioning?

5. Review your instructional materials. Identify one or two questions in an upcoming unit of study that you could revise to engage students in reasoning with conjectures.

4

Developing and Evaluating Mathematical Arguments

A good proof is one that also helps one understand the meaning of what is being proved: to see not only that it is true but also why it is true.

—Erna Yackel and Gila Hanna, "Reasoning and Proof," in *A Research Companion to Principles and Standards for School Mathematics*

The Role of Mathematical Arguments in Reasoning and Proof

Students in grades 6 through 8 are able to explain and justify the ideas they are reasoning about as they are problem solving. These justifications lead to students' ability to develop mathematical arguments, which become more elaborate than the informal justifications developed in earlier grades. Recall, for example, the proof presented for the *Mystery Number Riddle* in Chapter 2. In this dialogue, students developed a visual proof to reason about why the riddle always returned twice the original number. The proof consisted of demonstrating each step with a generic example rather than a specific value; students created an analytic (i.e., deductive) proof. Although middle grades students are not expected to write the type of formal proofs expected in high school (i.e., in a typical high school geometry course), their arguments should convey understanding of what is being proved and why it is true, as the opening quote to this chapter notes. They should also move beyond attempts at empirical proofs, in which they attempt to prove a statement using multiple examples (Sowder and Harel 1998). The reasoning used to develop mathematical arguments in the middle grades builds foundational understandings necessary for students to write formal proofs in high school.

Throughout the middle grades, students are continuing to learn how to describe generalizations and apply these descriptions to the formulation of conjectures. Students also must be expected to develop justifications and mathematical arguments to support or disprove their conjectures. Their mathematical arguments are being developed in a systematic way, and students are realizing that it is not necessary to test a conjecture endlessly to determine its validity. Students develop these justifications and mathematical arguments as they analyze relationships, properties, and mathematical structures that are the focus of their generalizations.

How Can You Help Your Students Create Justifications?

The daily discussions we have with our students must include opportunities for them to explain their reasoning. Help students verify whether an answer makes sense and decide if a strategy used in problem solving is efficient and why this strategy is efficient. These experiences help strengthen students' ability to self-monitor their thinking and reasoning. When explaining why a strategy used in reaching a solution makes sense, expect your students to give examples to justify their thinking. Students who are expected to explain and provide justifications are beginning the foundational skills necessary for the development of mathematical arguments.

We must consistently reinforce the expectation that students listen carefully to their classmates' reasoning so that they can determine whether an explanation of a conjecture is valid or whether there might be flaws in the student's thinking. This careful listening enables students to reflect on others' explanations and proposed conjectures and will enable them to identify errors in reasoning that later may become counterexamples. Remind students that when misconceptions occur in reasoning, they should be viewed as a valuable part of mathematical thinking and reasoning. When students are encouraged to examine flaws in reasoning, they begin to develop a deeper understanding of a concept or skill than when they simply complete a computation. This ability to reflect on and scrutinize reasoning helps students work through difficult problem-solving situations.

Planning experiences that include those just described provides students with opportunities to develop and evaluate explanations of justification and mathematical arguments. By always expecting students to explain why they know an answer is correct, why a problem-solving strategy always works, why a conjecture is true, or why a justification supports or disproves a conjecture, we give our students many ways to convince themselves and others that the reasoning shared is valid. These experiences prepare students to tackle more formal mathematical proof writing in high school.

Questioning That Promotes the Development of Mathematical Arguments

Just as questioning is critical in helping students learn how to make conjectures, so is it important to students' development of mathematical arguments. The questions we ask guide them in learning to think about the reasoning of others and their own reasoning as well. Students must become aware of the importance of justifications, which help them develop informal mathematical arguments. Students may be confused when you first ask questions such as, "If we try this with more problems, will it still work?" They are being asked to examine what they thought was a correct response. Through encouragement, support, and positive facilitation of discussions, we can help them recognize the benefits of searching for more justifications for why something works and makes sense.

CLASSROOM-TESTED TIP

Here are some questions that help students learn how to develop mathematical arguments:

- Why does this work? Will it work every time?

- If we try this with more problems, will it still work?

- Does this explanation make sense to you? Why or why not?

- What part of the explanation does not make sense to you? Why?

- What would happen if you . . . ?

- How will you convince us that this is true?

- Will this work with other numbers?

- Is this conjecture true if you consider other numbers (or shapes)?

- What do you think will happen if you solve ten more problems this way? Will the conjecture still be true?

- Do you think this strategy will work every time? Why do you think so?

Learning about the elements of a valid argument occurs as students participate daily in rich discussions about the mathematics they are learning. By establishing daily discussions as a part of your classroom routine, you offer students multiple opportunities to reason, notice generalizations, and make conjectures about a variety of mathematical ideas and relationships. Students will begin to understand that these generalizations and conjectures must be supported and justified by mathematical arguments. When students develop mathematical arguments, the mathematics knowledge they are learning becomes more explicit. This sense-making is crucial for our students' conceptual understanding of mathematics.

Helping Students Develop and Evaluate Mathematical Arguments

In Chapter 3, we presented several student dialogues in which students were making conjectures about properties of numbers, particularly the commutative and distributive properties. Although the purpose of the student dialogues was to engage students in the process of making or testing conjectures, the students also developed justifications and mathematical arguments as they thought about the conjectures and whether they were always true. The ability to develop more complete mathematical arguments grows over time. The discussion about the *Containers of Nuts* problem in Chapter 1 was an example of how students were beginning to develop informal mathematical arguments. In this problem, students devised various ways to organize the combinations to equal 12 pounds of nuts, and in doing so, they proved they had found all the combinations possible. You may want to review these dialogues from Chapters 1 and 3 before you continue reading.

In this chapter, we continue our discussion about what to expect as students communicate about and evaluate mathematical arguments. It is certainly not expected that students in grades 6 through 8 be able to develop and evaluate formal proofs about conjectures—this occurs in high school and beyond. But you can expect students to create justifications and informal mathematical arguments that are more developed than what they were able to do in earlier grades. They are beginning to understand that arguments can be developed from a more generalized thinking that is generated from classes of numbers and relationships.

Consider the student responses in Figures 4–1 and 4–2. Both students appear to know that the distance between two dots on the grids, either horizontally or vertically, is one unit. These responses contain justifications for how students obtained the area of a figure when a typical formula for finding the area was not known. In each case, students generalized a process to find the area—namely, divide the region into figures whose area can be determined. However, the student response in Figure 4–1 contains a number of misconceptions. First, the student seemed to consider a diagonal length from one dot to the other to have a length of 1 instead of $\sqrt{2}$, so the student obtained incorrect dimensions for each triangle. Second, the student divided the original region into three triangles and assumed they are congruent, which was an invalid assumption. Third, the student made a computation error ($2 \times 1 = 3$) in addition to using an incorrect formula for the area of a triangle ($A = bh$) before multiplying this result by 3 for three triangles. Although there are a number of errors in this student's response, the student's thinking and reasoning are visible for all to see. Therefore, the teacher is able to prepare instruction that will enhance the student's mathematical growth. If the student had simply provided an incorrect area without displaying his thinking, there would be little on which a teacher could build instruction to address those misconceptions.

In contrast, the student response in Figure 4–2 illustrates not only how the region was divided but how pieces were rearranged so that the area is the sum of a large square and a small square. Although minimal in words, the argument in Figure 4–2 is

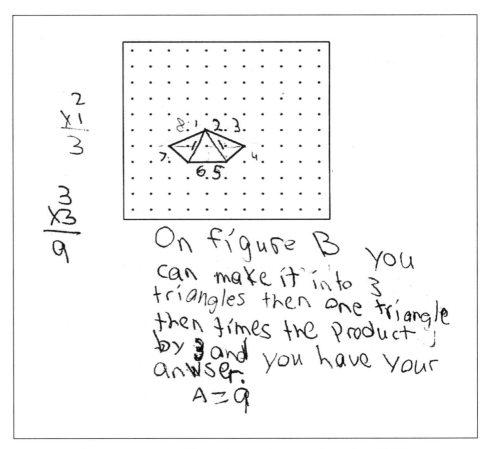

Figure 4–1 *Student argument illustrating a misconception when dividing a region to determine area.*

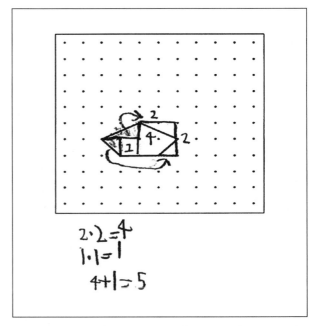

Figure 4–2 *Student response that provides a visual argument for finding the area of a region.*

clear. This response highlights the need for teachers to be willing to accept mathematical arguments in a variety of forms. As the opening quote indicates, the important aspect of an argument is that it conveys meaning and explains why something is true. This response clearly does that.

Encourage students to write about their mathematical arguments before presenting them to their classmates. When students record their justifications and mathematical arguments, they must first organize their thinking and reasoning in order to reflect on it. This helps them to reexamine their original arguments for any flaws and can spark additional reasoning that will enable them to refine their mathematical arguments. A written record also enables other students to critique and evaluate an argument and to learn from it.

CLASSROOM-TESTED TIP

Provide students with easy access to tools when developing mathematical arguments, including visual representations, manipulatives, and technology. Representing a mathematical argument with concrete materials, or a visual representation, is powerful. When the focus is on reasoning or problem solving, consider allowing students to use technology so that computational fluency, or lack thereof, does not impede students' reasoning ability.

Proportional reasoning is a major topic for study throughout the middle grades in all content domains. The *Strawberry Smoothie* task provides students with an opportunity to determine whether to reason from an additive perspective or a multiplicative perspective.

Carmelita loves to make smoothies on a hot summer day. Version 1 of her smoothie uses 2 cups of juice and $\frac{1}{2}$ cup of frozen strawberries. Version 2 uses $3\frac{1}{2}$ cups of juice and $\frac{3}{4}$ cup of frozen strawberries. Which smoothie version would have a stronger strawberry taste? How did you decide?

TEACHER: (*after students have had a chance to work on the problem on their own or with a partner*) So, which version do you think has a stronger strawberry taste?

STUDENT: We think Version 1 since there is less juice but the strawberry taste is all through it. If it is $3\frac{1}{2}$ cups with $\frac{3}{4}$ cups of strawberries it will have less taste since it has a lot of juice.

TEACHER: Let me restate what I think I hear you saying. You're saying that the version with the least amount of juice is the one that has the stronger strawberry taste. Will it always be the case that the version with the least amount of juice has the stronger taste?

STUDENT: Yes, because the strawberries get shared with less juice so it has to be a stronger taste.

TEACHER: Well, let's try something. Suppose Carmelita wants to make Version 1 of her smoothie for a party and decides she needs to use 7 cups of juice in order to have enough for everybody at the party. How many cups of frozen strawberries should she use to keep the strawberry taste the same? Take a few minutes to solve this problem. Be prepared to draw a representation or provide some argument to explain your thinking. (*Students are given some time to work on this problem before coming back to share responses.*) I'd like to have someone explain how they found an answer.

STUDENT: We decided to make a version 1 over and over and then took version 1 and cut it in half until we got 7 cups of juice. She needs $1\frac{3}{4}$ cups of strawberries. I can write our work on the board. (See Figure 4–3.)

TEACHER: How many of you agree with this group's thinking? (*About half the class nods yes.*) What about the rest of you?

STUDENT: I wanted to make a proportion but I couldn't get it to work. But I guess I can see what they did. So they just made groups of version 1 until they got 6 and when they needed 1 more they just took half of the version 1 amount. Now I get it.

TEACHER: I want to go back to our first discussion. Someone said Version 1 had the stronger taste because it had less juice. But now we have 7 cups of juice with $1\frac{3}{4}$ cups of strawberries and it is supposed to have the same flavor. So it seems we have a bit of a problem. We have more juice but the same flavor. So, it looks like we might need to revise our thinking about which one has the stronger taste or why it has the stronger taste?

STUDENT: So does that mean Version 2 has the stronger taste?

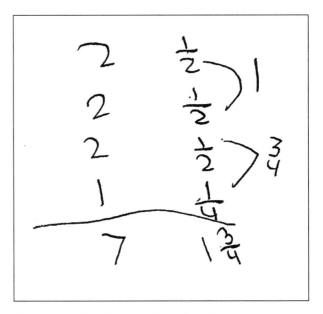

Figure 4–3 *Student work to show how to make a smoothie with 7 cups of juice but the same strawberry taste.*

STUDENT: No. I also think Version 1 is stronger strawberry but I did it a different way. You can't just look at the amount of juice. You have to compare the amount of juice and the amount of strawberries. You have to find the ratio of strawberries to juice and compare them. (See student's work in Figure 4–4.) Version 1 is stronger because the ratio is greater.

TEACHER: If she's right, what should the ratio be when Carmelita made smoothies using 7 cups of juice? I want someone to give me a prediction before you check it out.

STUDENT: It should be the same as Version 1 because we made it to be the same taste.

In this case, the first student's conclusion about the version with the stronger taste was correct, even though the reasoning to reach that conclusion was not. Rather than immediately seek another method to determine strength of flavor, the teacher chose to have students focus on a task that involved more juice but with the same flavor. In essence, the teacher helped students construct a counterexample to the argument that was provided. After students constructed a case with more juice but the same flavor, the teacher brought students' attention back to the original argument and helped them make revisions to focus on the ratio of strawberries to juice, thus engaging them in thinking about the task proportionally. The teacher then connected the revised argument with the work done on the intermediate task with 7 cups of juice in order to strengthen the concept of same flavors having the same ratio, regardless of the amount of juice.

It would have been easy for the teacher to have simply told students that the initial reasoning was incorrect or have told students how to make the comparison. Giving students the algorithm would have hindered their reasoning ability. Instead, students were allowed to grapple with the problem and to explore a case that discounted the argument. This process enabled students to revisit their thinking and provided an opportunity for some students to make connections to earlier work with ratios. Their explanation and demonstration of ratio comparisons was beneficial for their classmates' learning as well.

Figure 4–4 *Student response using ratios to compare strawberry flavors.*

Conditional statements (*if-then* statements) are important in developing formal mathematical arguments as students move through the grades. A simple activity that supports students' reasoning skills is one in which they are given *if-then* statements and must explain why the statement is true or find a counterexample to show why it is false.

- *If* it is a square, *then* it is a rectangle.

- *If* it is a trapezoid, *then* it is a quadrilateral.

- *If* a and b are factors of 16, *then* $a + b$ is a factor of 16.

- *If* a number is divided by $\frac{1}{2}$, *then* the quotient must get smaller.

In addition, students should be able to change statements in the form *All A are B* to a corresponding *if-then* statement and then determine whether it is true or not. Likewise, students should be able to explore statements of the form *Some A are B* and show how to determine whether the statement is true or false.

- *Some* triangles are equilateral.

- *Some* composite numbers have more than eight factors.

- *Some* prime numbers are even.

- *All* prisms have rectangular bases.

Statements can be generated from a variety of mathematical ideas or relationships. This classroom-tested tip was adapted from an activity in *Teaching Student-Centered Mathematics, Grades 3–5*, by Van de Walle and Lovin (2006).

In the Classroom-Tested Tip on page 52, we presented a list of questions that support students in developing mathematical arguments. Two of those questions were "Does the explanation make sense to you? Why or why not?" and "Do you think this strategy will work every time? Why do you think so?" Figures 4–5, 4–6, and 4–7 contain three different strategies to find the area of a figure on a grid, reflecting the following student dialogue.

TEACHER: I have three different solutions to finding the area of the figure on the grid. I want you to look at each solution and try to understand what the person was thinking. First, I want you to consider the accuracy of the reasoning. Second, I

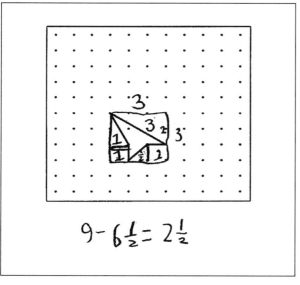

$$9 - 6\frac{1}{2} = 2\frac{1}{2}$$

Figure 4–5 *One student's method to find the area
of the figure.*

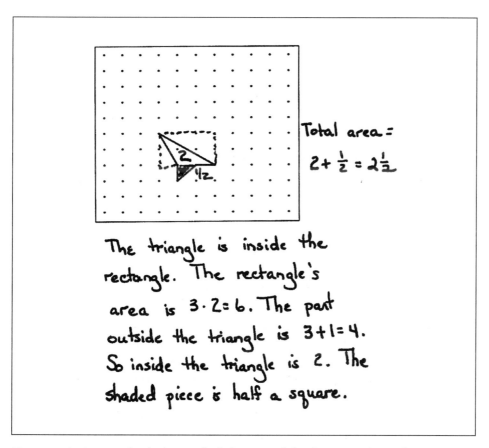

Total area =
$2 + \frac{1}{2} = 2\frac{1}{2}$

The triangle is inside the
rectangle. The rectangle's
area is 3·2=6. The part
outside the triangle is 3+1=4.
So inside the triangle is 2. The
shaded piece is half a square.

Figure 4–6 *A second solution to find the area of the figure on the grid.*

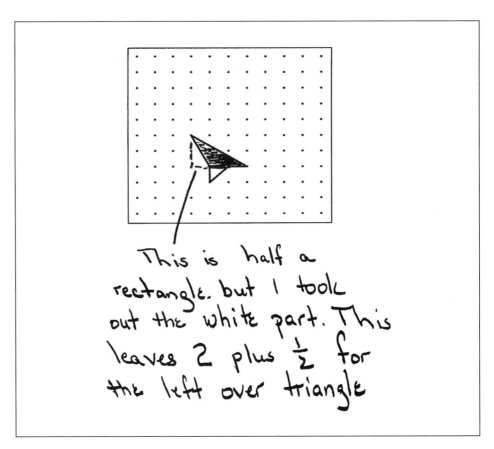

Figure 4–7 *A third solution to find the area of the figure on the grid.*

want you to decide if this method would always work. Third, I want you to decide how you would use this person's method to find the area of this new figure I am putting on the board. (See Figure 4–8.) Finally, I want you to decide which method you might like to use in the future and why. (*Students are given several minutes to investigate the solutions and share their thinking with the other members of their group.*)

STUDENT: The first student (Figure 4–5) made a big square around the figure and then subtracted out the areas from the parts that were outside the figure. All the outside parts were either triangles or rectangles, and it's easy to find the areas for these. I think you could always use this method because you could always put a square or a rectangle around something on the grid. But it might not always be easy, though, to find the areas of the unneeded parts outside the figure. If I used this method on the new figure, my work would look like this. (See Figure 4–9).

STUDENT: The second (Figure 4–6) and third (Figure 4–7) methods are almost the same. The second one puts a rectangle around the top part of the figure and just leaves the little triangle at the bottom out. Then the person finds the area of the part outside the figure just like the first person did. The third one just makes a right triangle and then takes out the unshaded part of the right triangle. I think the solution for these methods would kinda look like the one from the first person, except they might only put a rectangle around the top half of the figure because the bottom part is already a rectangle.

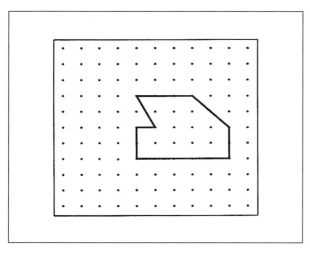

Figure 4–8 *Figure for which students are to apply
another person's strategy to find the area.*

The teacher's intent in presenting this problem to her students was to provide a situation in which they would be encouraged to understand an argument or solution
strategy made by another student. There are often multiple ways to approach a problem. In addition to reasoning about a problem using one's own approach, students' reasoning is strengthened as they evaluate strategies used by others to determine whether
they are valid or not. One way to assess whether students truly understand a valid approach used by another student is to ask them to apply the other person's approach to

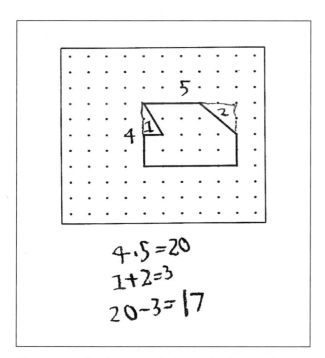

Figure 4–9 *Student use of the method from Figure
4–5 to find the area of the figure.*

a new problem. The teacher did this in the sample dialogue when she asked them to apply the method to find the area of the figure shown in Figure 4–8. We want our students to be able to reason about problems on their own and to evaluate and critique the reasoning of others.

Final Thoughts

The formal proofs that mathematicians formulate have rigorous standards. We have discussed the foundational understandings critical for students to develop so that they are ready to make more sophisticated proofs in higher mathematics. In grades 6 through 8, students' proof writing is accomplished in the form of justifications and informal mathematical arguments.

Reasoning is used "to navigate through the many facts, procedures, concepts, and solution methods and to see that they all fit together in some way, that they make sense. . . . Research suggests that students are able to display reasoning ability when three conditions are met: They have a sufficient knowledge base, the task is understandable and motivating, and the context is familiar and comfortable" (National Research Council 2001, 129–30). To understand mathematical ideas and relationships, students must use arguments to convince themselves and others that the reasoning is valid.

In grades 6 through 8, students learn that one counterexample is enough to demonstrate that a conjecture is not true. A conjecture may have many examples of why it is true, but if students do not examine a conjecture for the purpose of identifying a counterexample, they may develop flawed reasoning about the conjecture. Another example of the benefit for identifying a counterexample occurs when a conjecture is not clearly stated. This prompts students to state conjectures so that there is no possibility for a counterexample.

In Chapter 3, the teacher's students shared a counterexample when determining whether $r \div s = s \div r$ for all nonzero values of r and s. Even though they found some examples for which the statement was true, they continued to investigate the situation to determine if it was true for all values of the variable. Earlier in this chapter, the teacher helped students revise their reasoning about strawberry flavoring by engaging students in a task that served to counter the original argument and helped move students toward appropriate reasoning about the problem.

If students experience difficulties in making conjectures and developing mathematical arguments that support them, it may be because of their limited experiences in sharing explanations of reasoning. The classroom environment is a critical factor in providing a learning climate that encourages students to explain their reasoning with confidence. When students are unaccustomed to discussing and reasoning about mathematics, doing so is a challenge. Therefore, the establishment of a classroom climate that respects and nurtures our students' reasoning abilities and conveys the message that mathematics should make sense is a high-priority goal for us.

The role that traditionally describes teachers is that of the giver of all mathematical procedures and rules, with students being the recipients of this information. These traditional roles inhibit the development of students' reasoning abilities. In a standards-based instructional program, students are encouraged to explore mathematical ideas and the teacher becomes a facilitator of learning. Students who are engaged in meaningful mathematical learning routinely reason and make sense of what they are learning, notice generalizations, develop conjectures about mathematical ideas and relationships, and test their conjectures with justifications and informal mathematical arguments. These kinds of reasoning skills empower our students to become mathematical thinkers who understand that mathematics is about sense making.

Questions for Discussion

1. Reflect on or discuss with your colleagues: What is the importance of allowing students the opportunity to develop mathematical arguments?

2. How does the process of making mathematical arguments help students deepen their reasoning abilities?

3. What are some challenges that may occur as you begin to adjust your mathematics instruction to include (a) experiences for students that will support them in stating conjectures and (b) time to make mathematical arguments that will either support or disprove these conjectures?

4. Discuss with your colleagues: What are some actions you can take to make the challenges identified in question 3 more manageable?

5

How the Process Standards Support Reasoning and Proof

Learning with understanding can be further enhanced by classroom inter-actions, as students propose mathematical ideas and conjectures, learn to evaluate their own thinking and that of others, and develop mathematical reasoning skills.

—National Council of Teachers of Mathematics,
Principles and Standards for School Mathematics

The process standards described by the National Council of Teachers of Mathematics are the ways students learn and apply their mathematical content knowledge. We have discussed the process standard of reasoning and proof in detail. Now we will explore how the remaining process standards—problem solving, communication, representation, and connections—relate to and support reasoning and proof. The process standards are highly interconnected and support one another. The five process standards are integrated throughout our daily lessons as we provide experiences for our students to develop understandings of the five content standards—number and operations, algebra, geometry, measurement, and data analysis and probability.

Problem Solving

Problem solving and reasoning are highly dependent on one another. "Students who can both develop and carry out a plan to solve a mathematical problem are exhibiting knowledge that is much deeper *and* more useful than simply carrying out a computation" (NCTM 2000, 182). The way students think and reason about a problem-solving situation supports this deeper understanding and thus helps support their reasoning. NCTM has recommended that instructional programs from prekindergarten through grade 12 include experiences that enable students to learn new mathematics in a problem-solving manner. Problem solving should occur daily throughout our instruction, regardless of the content focus of the lesson.

NCTM has also recommended that students have multiple opportunities to solve problems in a variety of ways, so they may become flexible in how they think about mathematics. In today's classrooms, students should be able to self-monitor their problem-solving strategies for effectiveness. It is important that they reflect on why a strategy works for a particular problem, which strategy is more efficient than others for a range of problems, and how to refine a strategy that is not efficient.

Problem solving is the vehicle for learning mathematical knowledge. If students cannot make important decisions about solving problems, all the procedures and facts they have learned are of little use. The problems we present to our students must be engaging and challenging and provide opportunities for students to reason about mathematical ideas and relationships. We want students to understand that problem solving may take time, so we need to encourage them to be persistent problem solvers. The *Containers of Nuts* problem in Chapter 1 is an example of this type of problem solving. In the containers problem, students developed a variety of strategies in order to identify all the possible combinations for three containers so that the total quantity was 12 pounds with no two containers holding the same amount. Students were challenged to explain the reasoning they used to determine when they had identified all combinations. Interesting problems that allow students to grapple with the solutions both support and develop their reasoning abilities. We also must know when it is time to assist students who are struggling with a problem, because providing this assistance too soon will inhibit students from reasoning and making valuable mathematics revelations about their problem solving. Offering students rich and thought-provoking problems helps them become confident problem solvers who also use reasoning to solve and make sense of all kinds of mathematical situations.

Students often need to use tools in their mathematics study, including calculators. At times, our focus may be on developing computational proficiency with paper-and-pencil tasks or with mental strategies. In those cases, it is appropriate that calculators remain in the desk. However, when our focus is on problem-solving strategies and/or reasoning, students' lack of computational fluency must not inhibit their engagement with the mathematics; in these situations, technology enhances problem solving. In fact, calculators can be a useful tool to help students think about number sense, as the following dialogue illustrates. The dialogue is built on a series of related tasks, entitled *Multiplication Estimation*. (See Figures 5–1 and 5–2.)

TEACHER: Take a look at the first task on the board. Let's see how we might find a number to multiply by 3.2 so that our product is between 100 and 150. (*Students are given a minute to think about the task.*) Who would like to share what they were thinking and tell us why they chose that number?

STUDENT: Let's try 30. I know that 3 times 30 is 90 so I think 30 times 3.2 might be a little more than 100.

a. Multiply 3.2 by a number so that the product is between 100 and 150. Record the numbers you try each time. How did you decide what numbers to try?

b. Multiply 3.2 by a number so that the product is between 1000 and 1500. Explain how your work in (a) could help you do this problem in one step.

c. Multiply 0.32 by a number so that the product is between 1000 and 1500. Explain how your work in (a) or (b) could help you do this problem in one step.

Figure 5–1 *A series of three related questions to explore number sense from the* Multiplication Estimation *task.*

TEACHER: Let's try it. You put the problem in your calculator while I put it in the one here on the overhead. When I put in 3.2 times 30 I get 96 so it's not quite what we need.

STUDENT: It's pretty close. So I know I need something bigger. I know 3 times 50 is 150, so 50 would be too big. Can we try 40?

TEACHER: Try it on your calculator and tell me what you get.

STUDENT: 40 times 3.2 is 128, so it works.

TEACHER: What if we now want our product to be between 1000 and 1500? How can we use what we already know to find a multiplier for 3.2 so our product falls in this new range?

Figure 5–2 *Students exploring mathematics tasks with technology.*

STUDENT: We could just add a zero.

TEACHER: I'm not sure I understand what you mean. Can you clarify your thinking for us?

STUDENT: I just add a zero to the 40 to get 400 so 3.2 times 400 is 1280 and it works.

TEACHER: Let's talk about this a bit more. When you changed 40 to 400, you didn't really add a 0 to 40, did you? What really happened?

STUDENT: He multiplied 40 by 10, because when you multiply by 10 you just get an extra 0 on the end of the number.

TEACHER: Okay, so we were really multiplying by 10 using a shortcut. Why did you decide to multiply by 10?

STUDENT: The range for the numbers (*1000 to 1500*) was multiplied by 10 from the first problem. So I figured you needed to multiply the 40 by 10 as well.

TEACHER: What do the rest of you think about that reasoning? (*Students nod in agreement.*) What if we now wanted to stay in the same range but our starting number is 0.32? How can we use what we've already done so we don't have to start over?

STUDENT: We divided 3.2 by 10 so I think we need to divide 400 by 10. So I think we can use 40 like we did before.

STUDENT: I tried that on my calculator and got 12.8, which doesn't work. Then I decided to multiply 0.32 by 400 and got 128, which still doesn't work. So I finally multiplied 0.32 by 4000 and got 1280. I'm trying to figure out why it worked that way.

TEACHER: We have an idea on the floor. Take a minute and talk with your partner. Why did we need to multiply by 4000 when we changed our number from 3.2 to 0.32? (*Students have time to talk about the problem with their peers before coming back together for a discussion about the inverse relationship of multiplication and division.*)

Students were using deductive reasoning when they recognized that if the range for the product increased by a factor of 10 then the multiplier also needed to increase by a factor of 10. Opportunities for deductive reasoning also occurred when students recognized that if the product of two numbers is constant and one decreases by a factor of 10 then the other must increase by a factor of 10 (the relationship that existed with 0.32×4000).

There are many opportunities for using number sense to engage students with problem solving while making connections to other content areas. Consider the following relatively straightforward task, titled *Sums and Products,* on the CD:

Find two integers whose sum is 75 and whose product is 1400. Explain why the numbers are correct.

Again, this is a good task for the use of calculators so that computational fluency is not a hindrance to the thinking that we want students to use. Figures 5–3 and 5–4 contain two sample responses to this task. Both students provide an argument to show why their numbers work and how they made their decisions. The student whose response is in Figure 5–3 started with two numbers that add to 75; when their product was not the desired result, he tried additional numbers. But because their sum was constant, he recognized that when he increased one of the numbers he had to decrease the other by the same amount. The student whose response is in Figure 5–4 appears to have done something similar. Although the first student's argument suggests a systematic checking of numbers, the second response suggests a more random approach provided that the numbers sum to 75. The student uses the term *square root* to indicate the starting numbers. Even though taking the square root of 1400 would be one way to start, discussion with the student indicated she really took half of 75 as a starting point. Encouraging students to explain their reasoning provides opportunities for them to use vocabulary and articulate their thinking. Students' reasoning explanations present us with insight into potential misuses of important language, which can then be corrected before they pose difficulties later.

On the surface, the task represented by the responses in Figures 5–3 and 5–4 seems to be only a number sense task. But in algebra, students might use x and y for the two numbers and represent this problem as the system, $x + y = 75$ and $xy = 1400$. Graphing the two equations yields a line and a hyperbola, respectively, and the solution is the coordinates of the intersection of the two. So, exploring this and similar number tasks within a problem-solving and reasoning perspective lays a foundation for later work in algebra.

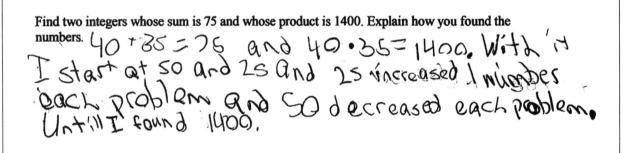

Figure 5-3 *One student's response to the* Sums and Products *task.*

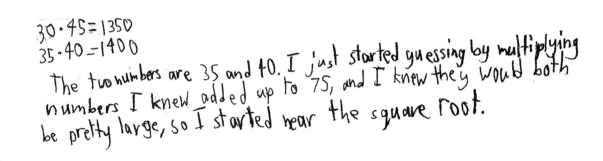

Figure 5-4 *A second student's response to the task in Figure 5–3.*

Below, we summarize some of the benefits of activities like those just illustrated:

- Students were engaged in deductive reasoning.

- Students were reinforcing their knowledge of number sense, specifically their understanding about how numbers relate to each other and how a change in one number influences a change in another.

- The teacher was obtaining important information about what her students knew about number relationships, how they were reasoning, and their ability to explain their thinking in writing.

The activities provided on the CD are problems that will be both engaging and thought-provoking for your students. They will help your students strengthen their number sense, problem-solving, and reasoning abilities. Whether you allow your students to complete these problems as a whole group or independently, the discussions you facilitate with students are critical. If students will sometimes be completing the activities independently, it is important to bring students back together to discuss their problem-solving and reasoning strategies. These discussions give students opportunities to share multiple strategies and talk about the reasoning they are using in solving the problems. And you will learn a lot about your students' understandings!

In the beginning, students may begin by sharing more procedural strategies as they explain the different ways they solved the problems. This may be particularly true if much of their prior mathematics experience has focused on algorithms. In addition to focusing on how a problem was solved, always expect your students to discuss why the strategy used in problem solving makes sense. Gradually you will want to guide the discussions in such a way that you are supporting students in developing mathematical arguments that support the conjectures they are learning to formulate. The discussions you have with your students will be a source for informal assessments of how students are reasoning about mathematical ideas and relationships.

Communication

The process standard of communication plays a critical role in supporting reasoning and proof. Talking and writing are the ways that students make their thinking about mathematics visible and explicit. NCTM has recommended that instructional programs from prekindergarten through grade 12 include experiences that enable students to communicate the mathematics they are learning in a variety of ways. In order for students to communicate their thinking and reasoning effectively, they must first organize their thinking. This is also true when students communicate in writing. In grades 6 through 8, it is crucial for students to communicate ideas and reasoning so that others clearly understand their thinking. Mathematical language must be used correctly to describe the ideas and relationships that are the focus of students' reasoning. We previously discussed the value of students' ability to self-evaluate their own thinking and reasoning, but there is also a need for students to evaluate the reasoning of others. The ability to evaluate others' reasoning allows students to reflect on mathematics, which in turn helps them to make sense of it. In Chapter 4, we provided dialogues of students attempting to understand the reasoning strategies used by others for area problems on a grid.

Reasoning and proof are interconnected and highly dependent upon the communication standard. This relationship has been demonstrated in the student dialogue examples throughout this book. In these dialogues, students have communicated in collaborative groups in order to problem solve together and have shared their discussions in whole-class settings. When students are presented opportunities to talk about mathematics, it helps them to reflect on what they are learning and to make sense of it. In addition, the discussions we facilitate with our students provide us with important information about what our students understand about the mathematics we are teaching them. It is easier to recognize a student's misconception when it is revealed in a student discussion rather than left unspoken in a classroom where discourse is not an expectation. The knowledge you gain about how your students are reasoning and making sense of mathematics provides you with valuable information that allows you to plan instruction that meets the needs of all students.

CLASSROOM-TESTED TIP

There are several ways to organize your students to enhance student discourse. These formats should vary according to the needs of your students and your instructional goals for the lesson.

- A whole-group discussion is demonstrated in many of the student dialogues presented in this book. During these discussions, the teacher acted as a facilitator to guide students' thinking and reasoning. Instead of the focus being only on the answer, the focus was on how students were thinking and reasoning about the process used to reach the solution. These whole-group discussions provide students with essential experiences that will support their development of reasoning skills.

- Small groups are utilized when students have begun in a whole-group discussion but there is a need to move into smaller groups in order to investigate the mathematics further. In several of the student dialogues demonstrated, students began as a whole group and then formed smaller groups to discuss the task. In small-group discussions, the teacher circulates among the groups as students formulate reasoning about a mathematical idea. Instead of controlling these discussions, our role should be that of an observer. As an observer, we can assist students' learning by redirecting their thinking if necessary. These small-group discussions will provide us with a way to assess how our students are reasoning informally.

■ The last format for discourse is partner talk. In this format, the teacher asks a question and provides partners a short amount of time to discuss it together. This format is usually implemented within a whole-group discussion. If you find that wait time is not producing more student talk in a whole-group discussion, a change to partner talk might help. Partner talk is especially beneficial for students who are not comfortable talking in front of a larger group. It gives students time to talk in a less threatening environment. If a student is confused, he'll be more likely to talk about this confusion with a peer. Students who are learning to speak English find that partner talk helps them to adjust to their new situation. The short time that you provide for students to talk with a partner prepares all your students to be active participants in a whole-group discussion.

In the sample student dialogues provided in the various chapters, we have demonstrated students' reasoning about a variety of mathematics topics. In these experiences, students were given wait time. This strategy allowed them an opportunity to reflect on their thinking and thus encouraged reasoning. When students worked in small groups or talked with a partner, these experiences were also helping them to organize and hear different interpretations of reasoning within their groups. Students were learning to clarify their reasoning, conjectures, and mathematical arguments when they were asked to restate their own thinking or the thinking of others. In the student dialogues, mathematical language was being modeled by the teachers and the students. Specific questions that encouraged student reasoning were modeled by the teachers.

It is important to point out that the student discourse occurring in these classrooms was a result of established routines. The ability for students to discuss mathematics meaningfully develops over time. In Chapter 1, we shared important factors that you should consider when establishing student discourse as a routine in your classroom. One of the most important factors that directly affects the success of student discourse is the climate and environment of a classroom. We establish the climate at the beginning of each school year, and our goal should be to create a community that allows students to feel safe and comfortable. Our students will be more likely to take risks and share reasoning in this type of a learning community than in one that feels competitive or unsafe. Students will not feel threatened if their reasoning contains flaws, and the opportunities to examine and analyze these misconceptions will further develop your students' mathematics learning.

It is important that students have an understanding of how your classroom environment and routines are organized. In order for student discourse to become a productive routine, we must clearly explain our expectations about the students' role in these discussions. You may find that, for some students, participating in student discussions will be a new and perhaps intimidating experience. When students know and understand what is expected of them, it provides them with a rationale for why and how they are to communicate reasoning in your classroom discussions.

At the beginning of a school year, it may be helpful to have sample responses from previous years or from other classes when students begin to evaluate reasoning explanations. Students will be critiquing generic responses, which helps them to understand the importance of evaluating others' reasoning while providing them with a model for constructive critiques.

CLASSROOM-TESTED TIP

Here are several expectations that are important for your students to understand about their participation in student discussions:

- Students will communicate their thinking to you and to other students.

- Students will question you and other students in order to help them clarify their own thinking.

- Students must explain and justify their thinking and reasoning.

- Students will view flawed reasoning or misconceptions as learning opportunities.

- Students will understand that everyone's ideas are valued and that the evaluation of others' reasoning is done in a respectful way.

Many teachers find it difficult to transition from a classroom of more teacher talk to one of more student talk. As you begin to encourage and support more student talk, you may hear students sharing thinking and reasoning that is flawed or includes misconceptions. Teachers often find it difficult to know how to respond to a student's flawed reasoning. Understanding the mathematics we are teaching students will help. To help our students sort through their misconceptions, we need an awareness that goes beyond the basic skills, rules, and procedures of mathematics. It is critical that we understand how students develop conceptual understanding of the mathematical concepts and skills we are teaching. This deeper understanding of the mathematics we are teaching will support us in facilitating student discussion in which flawed reasoning occurs.

Planning with colleagues in grade-level team meetings provides you with opportunities to discuss how students conceptually develop understandings about mathematics. Discussion among colleagues across grades provides insights into how understandings grow as students experience additional mathematical concepts. Such cross-grade discussions also provide insights into how students at different ability levels might respond. For example, advanced sixth-grade students might respond similar to lower-ability eighth-grade students. Conceptual understandings have more depth than that of a memorized rule or procedure. Identify the prerequisites of the concepts and skills that students will learn when studying a particular topic. This helps you to determine whether students have sufficient understandings about the mathematics to be learned. As you informally observe students, you can determine if a student has gaps in her learning about a concept or skill. Note these informal observations to help when planning with your team.

Discuss how your state or district grade-level objectives develop understandings for concepts and skills that will be learned in later grades. These planning times with your grade-level colleagues will help guide the facilitation of discussions you have with students. When students demonstrate they are missing key understandings necessary for the objectives or students contribute information that is beyond what is being taught, you will be prepared to address these issues.

When beginning any new topic in mathematics, discuss with your grade-level team any misconceptions you think your students may have about the mathematics. Anticipating potential misconceptions helps you plan your lessons. You will be better able to redirect students' thinking and reasoning if you have an understanding of what these misconceptions could be.

Even though we have emphasized how important it is to have less teacher talk, our role remains essential in the student discussions we facilitate. We must ensure that our student discussions flow in a focused manner and that we ask probing questions to promote reasoning, which will enable us to delve deeper into how our students are thinking. We must convey a message to our students that mathematics is about sense making, and students should know that their thinking and reasoning about mathematics are valued.

In addition to discourse, writing is also an effective communication strategy for supporting reasoning and proof. When students write about their thinking and reasoning, they must reflect, clarify, and then organize their thinking to write about it. Their writing is a record of their thinking, and it becomes a tool that helps them to analyze and reflect on the mathematics as they are writing about it. Writing is a way for students to internalize ideas about mathematics (Countryman 1992). Equally important, our students' writing provides us with another glimpse of how they are reasoning and reveals the understandings or misconceptions they may be having about mathematical ideas and relationships.

C L A S S R O O M - T E S T E D   T I P

Before students can be expected to write about mathematics, they must first talk about mathematics. Students benefit from multiple experiences in which they are talking about mathematical ideas and strategies. These experiences provide the foundation needed to write meaningfully about mathematics.

Math journals are often used to encourage students to write about the ideas they are learning in mathematics. They are beneficial for students because they provide a chronological record of their mathematical learning during the school year. In order for students to write about mathematics, it is necessary for them to reflect on the mathematics from the activity, which then helps them organize their thinking. Students will be thinking about what was done with the mathematics during the lesson and what was required to complete the problem solving in the activity. Math journal writing is effective when it occurs after a student discussion because students have had an opportunity to talk about the mathematics they will be writing about in their journals. (See Figure 5–5.) It is often difficult for students to write about something they have not discussed.

Math journals can also be used at the beginning of a lesson. Ask students to write what they already know about the concept that is the focus of that day's lesson, or what they think they will learn about the concept and why they think this. Another use of journals is to help you determine what the students understand about a specific concept or skill. Ask them to select what was the most difficult part of the mathematics lesson and to explain how they reasoned about it. Near the end of a unit, students can be asked to write about what they understand well and what is still causing difficulty; your review can then be tailored to what students perceive their needs to be. In addition, you are helping students build their self-monitoring skills, which are essential to success in mathematics.

The feedback we provide to students in their mathematics journals should be specific and helpful to their continued mathematics learning. Feedback that is vague tells students little about their thinking and reasoning. Specific feedback that is precise and helps students to reflect further on the mathematics they are learning is vital. Simply writing *Good Job!* or *Great* or *Super* on a student's journal entry does not provide them with useful feedback about their thinking and reasoning. Vague feedback also discourages students from further reflecting on their reasoning. Whenever a student's writing response includes flawed reasoning, our responses must redirect their thinking so that they will reflect on and analyze the flaws. This process will enable them to develop a clearer understanding of their original reasoning.

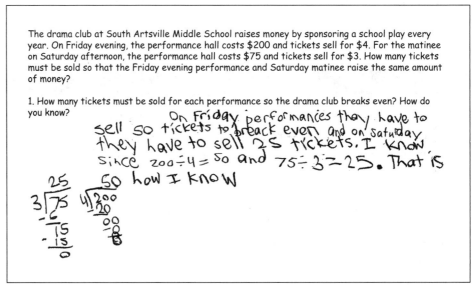

The drama club at South Artsville Middle School raises money by sponsoring a school play every year. On Friday evening, the performance hall costs $200 and tickets sell for $4. For the matinee on Saturday afternoon, the performance hall costs $75 and tickets sell for $3. How many tickets must be sold so that the Friday evening performance and Saturday matinee raise the same amount of money?

1. How many tickets must be sold for each performance so the drama club breaks even? How do you know?

On Friday performances they have to sell 50 tickets to break even and on Saturday they have to sell 25 tickets. I know since 200÷4=50 and 75÷3=25. That is how I know

Figure 5-5 *Student explanation of how a response to a problem was determined.*

Mathematics journals do not have to be collected daily; rather, collecting journals once a week so that you can also stagger which class periods submit journals on a particular day helps you manage the time necessary to respond specifically to students' journal entries. These writing assignments are informative evidence of what our students understand about a particular mathematics idea at that point in time. They are visible evidence of mathematical learning that we can also share with parents and that can be included in student portfolios.

An alternative to a formal mathematics journal writing activity is to have students write an exit slip at the end of the mathematics class (Daniels and Zemelman 2004). This can be a short writing assignment in which students indicate what is still puzzling them. These short writing assignments (or assessments) do not take a long time for teachers to read, yet they provide valuable information to guide further instruction.

CLASSROOM-TESTED TIP

Suggestions for mathematics journal prompts:

- What was easy in the mathematics we did today? Why?

- What was hard in the mathematics we did today? Why?

- What questions do you still have about the mathematics we did today?

- What suggestions would you give a friend to solve this problem?

- Explain why your way of solving the problem makes sense. Solve it another way and describe why the other way works, too.

- Finish this sentence: I wish I knew more about . . .

- Explain why the conjecture we discussed today is always true.

- Our proposed conjecture is _____. Do you think this is always true? Explain your reasoning.

- How is _____ like _____? How is it different?

Another motivating writing format that students enjoy is writing a letter to an absent student about the mathematics learned from that day's lesson. Students write about the reasoning discussed during the lesson's problem-solving activity or explain what they understand about the mathematics in the activity. Students' letters can be a type of informal assessment for the teacher. Letters allow you to monitor their understandings about the concepts and skills they are learning, which will guide you in planning instruction. Of course, letters should also be shared with the absent student! This writing activity then becomes purposeful, and it helps students understand the importance of describing mathematical thinking and reasoning clearly so that others will understand it.

CLASSROOM-TESTED TIP

Here are several strategies for supporting students in writing about how they are thinking and reasoning:

- Talking helps. If a large-group discussion is not possible first, ask students to discuss a task with a partner or in a small group before beginning a writing assignment.

- Occasionally ask pairs of students to think about a task together and then write a combined statement of reasoning.

- Before they submit a writing assignment, expect students to review and revise their explanation of reasoning.

- Invite students to share their written explanations of reasoning. This helps them to identify any ideas that are not clearly explained.

Opportunities for students to write about mathematical understandings are provided in the various CD activities. In many of these activities, students write to explain their reasoning in solving a problem. These writing experiences benefit students because they are writing to convince the reader that their reasoning in the problem solving is correct and to explain why it is correct. These writing responses will provide you with informal assessment data about how your students are solving and reasoning about a variety of tasks. Reading their responses will provide you with opportunities to ask students questions about how they thought about problems in order to solve them. When students' written responses reveal misconceptions or flaws in their reasoning, it will give you information you can use to help these students.

Representation

"It is important to challenge and encourage students to connect and to compare and contrast the utility and power of different representations" (Reys et al. 2004, 107). The process standard of representation supports students' reasoning in several ways. The models or representations that students develop provide a connection between a mathematics concept and the symbols used to explain that concept. Students will be applying their concrete learning to more abstract aspects of mathematics. Allowing students to choose how to represent their thinking and reasoning about a mathematical idea helps them to make sense of the mathematics they are learning. In Chapter 2 (Figure 2–5), some students represented the proof for the *Mystery Number Riddle* using boxes for the variable and dots for the numbers. Students in another class (see Figure 5–6) preferred to represent the proof using algebraic variables. It is important for teachers to be sensitive to students' different representations of the same task, which are based on their comfort level and prior experiences.

We want students to be able to reason from tasks expressed in multiple representations. It is important that students have opportunities to reason about problem-solving situations that are represented graphically. (See Figure 5–7.) In this particular task, all the pertinent information is contained within the graph. Students need to reason from the general problem description to determine what portions of the graph might represent shopping, lunch, or a visit with grandmother. In addition, students need to use the units on the axes to recognize that the slope of the segments represents speed; the steeper the slope, the greater the speed.

As students reason from representations, they have additional opportunities to communicate about their understanding. The student's story (see Figure 5–8) about the graph is cryptic but essential information about the ability to reason from the graph is evident. Although it seems clear that the student is able to interpret the graph, the student did make an error in describing the time Ruby spent visiting her grandmother. She read the 4 on the horizontal axis as the total time spent visiting her grandmother. However, the distance was constant (indicating no movement) from a time of 1 hour to 4 hours, so the visit had a duration of only three hours. Given that the student interpreted the other horizontal segments correctly, we might assume that the student was careless when describing this portion of the graph. A discussion with the student about the story would be in order.

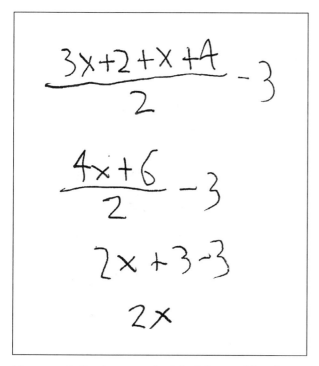

Figure 5–6 *Student proof of the* Mystery Number Riddle *task from Chapter 2 using variables.*

The graph at the right shows Ruby's distance from home at different times during a trip to visit her grandmother. During the trip she went shopping at an outlet mall and had a leisurely lunch.

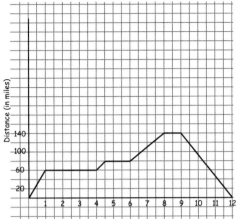

1. What is the farthest distance Ruby traveled from home? How do you know?

2. Which parts of the graph could represent Ruby's shopping time at the outlet mall, her leisurely lunch, and her visit with her grandmother? Explain your thinking.

3. When was Ruby's speed the fastest? When was it the slowest? Explain why your answers are correct.

Figure 5–7 A Day on the Road *task that requires students to reason from a graphical representation.*

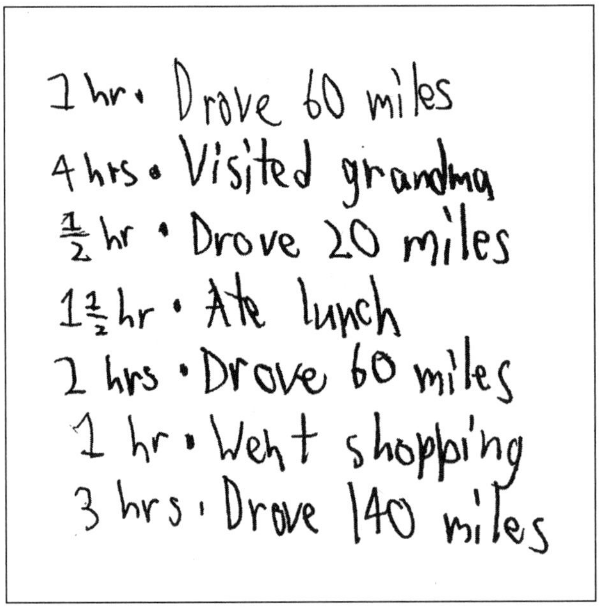

1 hr. Drove 60 miles
4 hrs. Visited grandma
½ hr. Drove 20 miles
1½ hr. Ate lunch
2 hrs. Drove 60 miles
1 hr. Went shopping
3 hrs. Drove 140 miles

Figure 5–8 *Student story to describe the graph in Figure 5–7.*

Students' representations demonstrate their understandings about mathematical relationships and reveal their reasoning to us. Students utilize representations as tools to problem solve, communicate thinking, and express mathematical ideas, and their representations are used to support or disprove conjectures. Representing thinking enables students to organize and make sense of their thinking and reasoning. NCTM has recommended that instructional programs in prekindergarten through grade 12 include experiences that enable students to make a variety of representations to help them think and reason about mathematics. Throughout the middle grades, students should be encouraged to use graphs, tables, symbols, and numbers to connect ideas. Students in grades 6 through 8 are capable of choosing which representation is appropriate for a mathematical situation and must be encouraged to be flexible in which representation to choose.

"Good representations fulfill a dual role: they are tools for thinking and instruments for communicating" (NCTM 2000, 206). When students represent their reasoning, it affords them opportunities to visualize their thinking. A student's initial reasoning may be incorrect, but this flawed reasoning might not be apparent to the student if he has not had the opportunity to view the reasoning in a representation. Students' representations extend their thinking and are a method to justify reasoning to others. Throughout the middle grades, students are learning that algebraic and numeric representations are useful for showing reasoning about relationships. At times, these relationships can be illustrated via graphic displays (e.g., line graphs, tables, circle graphs). Technology, particularly graphing calculator technology, supports students as they integrate numeric, tabular, and graphical displays to reason from one representation to another. Students should understand that representations are more than just *showing* how a problem is solved; representations are tools that help students develop understandings about mathematical ideas and communicate information about thinking and reasoning in an organized manner.

As mentioned earlier, we want students to understand that some representations are more useful than others. We can support our students in determining the most appropriate way to represent their reasoning by presenting different representations and discussing with them which representations are more effective in describing and justifying their reasoning. When students are first learning to represent a conjecture they do so in words. Eventually they will find that a conjecture is more precisely stated in a symbolic representation. Students should become comfortable using a variety of representations so they are able to choose those that are more useful for reasoning about different mathematical situations.

Connections

As students develop reasoning about mathematical ideas and relationships, they are forming connections from previously learned knowledge in a variety of areas. "These connections help students see mathematics as a unified body of knowledge rather than as a set of complex and disjoint concepts, procedures, and processes" (NCTM 2000, 200). In the connections process standard, NCTM has recommended that instructional programs in prekindergarten through grade 12 include experiences that enable students to see how a mathematics concept relates to other mathematical ideas, how these ideas overlap one another and build on one another, how mathematics is connected to other disciplines such as social studies or science, and how mathematics is connected to their lives outside of school.

It is critical that we provide experiences for our students that will enable them to appreciate the rich connectedness of mathematics. Students in grades 6 through 8 should be reasoning about mathematical ideas and relationships found in real-world situations, in mathematics lessons, and across other disciplines. The reasoning that students connect to their everyday lives will prepare them for making decisions in future years. When students see connections across mathematical concepts, it helps them understand that mathematics consists of closely connected ideas that support one another.

It may be necessary to plan ahead in order to highlight mathematics where appropriate. If your middle school is organized in teams, talk with your colleagues from other disciplines about what they are studying at a particular point in time so that you can make connections to those disciplines when possible. Focus on mathematics that is useful for students to apply in their everyday lives. Science is a discipline in which mathematics is highly integrated with reasoning, which is a critical part of any scientific investigation as students collect data from experiments, display the data, and make inferences from it. Disciplines such as social studies rely on mathematics when students determine distances on a map by reading and interpreting a scale correctly. These practical applications of mathematics may even stop students from asking, "When will I ever use this math?"

Our role is to help our students make connections by guiding them to see these connections in a variety of contexts and models. The connections our students are able to make with our support will help them refine their reasoning in a broader sense. A very simple way to approach this is to encourage students to look for ways in which mathematics is used throughout the school day, at home, or in their community. When students apply reasoning to everyday contexts, it provides them with a rationale for why mathematics should make sense and why it is important. Consider the following task, *Lots and Lots of Printer Cartridges*, a context that should be familiar to many students.

For a small desk jet printer, a black cartridge prints 450 pages and a tricolor cartridge prints 400 pages. Cartridges can be purchased in the following packages:

- 1 black $19.99
- 1 tricolor $34.99
- 1 double package with 2 black cartridges $35.99
- 1 double package with 2 tricolor cartridges $62.99
- 1 combo package with 1 black and 1 tricolor cartridge $49.99

Donald is working on a project that will consist of about 3200 pages. About one-third of the pages will have color on them. Donald wants to make sure he has enough cartridges for the job before he begins printing. What would

be the best cartridge package for Donald to buy so that he will be able to print the pages needed for this project? Use words, pictures, or numbers to explain your reasoning.

When students reasoned about this problem (See Figure 5–9), they first determined how many color or black cartridges would be needed. Some students constructed a table to check prices for all the different arrangements. However, the student whose response is in Figure 5–9 quickly did some mental mathematics to reason that the combo or double packages were cheaper than the corresponding number of individual packages. She then reasoned that to get 3 color and 5 black cartridges the best purchase would be three combo packages and then one double package for the remaining two black cartridges. This student used mental estimation and then applied reasoning in an authentic context. We want students to have this flexibility in their thinking.

Utilizing children's literature is an excellent strategy to help students make connections in mathematics, even in the middle grades. Math-related literature books are engaging and present students with problem-solving situations that reflect real-world mathematics. The use of literature in mathematics instruction enhances students' learning and reasoning by

- providing a meaningful context for the mathematics being learned;

- supporting relevant problem solving;

- integrating mathematics into other disciplines;

$$\frac{3200}{3} = 1066.\overline{6} \approx 1067$$

$$3200 - 1066.\overline{6} = 2133.\overline{3} \approx 2134$$

$$\frac{1067}{400} = 2.6675 \approx 3$$

$$\frac{2134}{450} = 4.74\overline{2} \approx 5$$

Donald should buy three combo packages and 1 black double package.

Figure 5–9 *Student reasoning about the* Lots and Lots of Printer Cartridges *task.*

▓ relating mathematics to everyday situations; and

▓ engaging students in rich problem solving that strengthens their number sense.

Children's literature easily becomes a springboard for further investigating the concepts or skills embedded in a story. The *Harry Potter* series of books and movies such as *Lord of the Rings* are resources from which rich mathematics investigations can be developed (see, for example, Beckmann, Thompson, and Austin 2004). These investigations support reasoning about mathematics that is connected to students' lives and help them understand the relevance of the mathematics they are learning. Several resources about how to integrate children's literature effectively into middle grades mathematics lessons are included in the Teacher Resources at the end of this book.

Final Thoughts

The process standards—reasoning and proof, problem solving, communication, representation, and connections—are the processes by which students acquire and use the mathematical concepts and skills they are learning. They are interconnected and interdependent on one another. When teaching in the manner that NCTM envisions for our students, we must consider the process standards as well as the content standards when we plan our instructional programs. Reasoning and proof is considered to be one of the most important NCTM standards; however, reasoning relies on the remaining process standards to support its development and effective delivery. Rich problem-solving tasks are the vehicle for promoting the reasoning skills we want our students to develop; communication and representation are how they reveal their reasoning to us and others; and the connections they make among mathematical ideas, across other disciplines, and to their everyday lives will support them in reasoning about mathematics in a variety of settings.

Questions for Discussion

1. How do the other process standards help support the reasoning and proof process standard?

2. Reflect on the interconnectedness that the process standards have with each other.

3. Which of the remaining process standards do you feel will be the most challenging for you to implement into your classroom teaching as you help your students develop their reasoning skills? Why do you think so? Discuss with your colleagues a way to overcome this challenge.

4. Talk with members of your team who are teaching other disciplines. Find at least one connection with mathematics for each of the other disciplines students study.

CHAPTER 6

Assessing Students' Reasoning

Knowing students' perceptions and misperceptions of a concept is key to developing their mathematical reasoning.

—Jillian N. Lederhouse, "The Power of One-on-One"

Why Assess Students' Reasoning?

It is important that we know how our students reason. The reasoning they develop supports them in making sense of the mathematics they are learning. We should make it a priority to unlock the mysteries of mathematics for our students by providing them with multiple opportunities to reason about it. Reasoning skills are critical to our students' everyday lives now and in their future endeavors. When we encourage our students to share and discuss reasoning, we open a window that presents us with a view of their thinking and understanding about important mathematical ideas and relationships.

How, then, do we measure the development of our students' reasoning abilities? It sounds difficult, and in this chapter we share strategies to make assessment of reasoning a manageable task. The first step, and one that is a critical part of the assessment process, is acknowledging that assessment is an integral part of your mathematics instruction. When assessment is an ongoing process, it informs and directs the decisions you make for all students.

A variety of assessments are used in schools today. In many of these assessments, the focus is generally on what students do not know. Common types of assessments are paper-and-pencil tasks, which generally measure a student's ability to complete a procedure. On this type of an assessment it may be difficult to analyze a student's error if no explanation of thinking or reasoning is offered. Administering only this type of assessment gives us an incomplete picture of what our students know. Although there is value in administering formal assessments, we must also make it a continual goal to seek information about our students' progress by using informal assessments. Assessment of students' reasoning is typically not easy when using traditional tests that are given in a timed, on-demand format. To assess how our students are reasoning about mathematics, we need to use formative assessments or use summative assessments over periods of time longer than a class period.

The importance of ongoing assessments that measure the continuing growth of students' mathematical understandings has already been stressed. This is also true of their reasoning skills. Making assessment a part of daily instruction gives us a broader method to determine whether our students have a deep understanding of the mathematics being taught than simply using paper-and-pencil tasks at somewhat regular intervals. When we assess students' reasoning, the process does not look different from instruction; rather, it should become a large part of our instruction. It is far more advantageous to focus our assessments on what students know about the mathematics concepts and skills they are learning and how they are reasoning about them than to focus only on what they do not know.

Formative assessments, which involve authentic contexts in which students are assessed in a variety of situations, effectively measure students' development of reasoning skills. These assessments include interviews, mathematics journals, evidence of students' self-monitoring, performance tasks, and observations. Students are assessed as they are meaningfully engaged in mathematics.

Students' explanations of reasoning are also revealed when they represent and communicate their mathematical thinking, which is one way to make assessment an ongoing and integral part of your daily instruction. Opportunities to assess reasoning can easily occur in the daily discussions you have with your students. Students' explanations that describe a variety of mathematical ideas and relationships provide you with important information about what they know and understand about the mathematics you are teaching. This chapter explains the usefulness of a variety of formative assessments that can help you assess students' reasoning skills and can be easily implemented into your classroom routine. It may seem difficult for teachers in grades 6 through 8, who often have five different groups of students during the day, to talk to students or find ways to assess them on an ongoing basis. However, with careful planning and a commitment to gauge students' reasoning skills, we can find a few minutes each week to talk individually with each of our students.

How to Assess Students' Reasoning

A variety of strategies support the assessment of students' reasoning. These assessment strategies differ from traditional assessments and allow you to do more than simply check off a correct or incorrect answer. The assessments described here are informal assessments that are integrated into daily ongoing instruction and will help you measure the growth of your students' reasoning skills. We begin by describing the classroom environment that allows you to do this.

NCTM recommends that teachers develop their classrooms into problem-based environments in which students learn mathematics by problem solving. This type of teaching focuses students' learning on mathematical ideas and sense making. In mathematics that is taught traditionally, students tend to focus only on the directions or the procedures that are presented to them rather than on the more meaningful ideas of mathematics. The value of teaching by problem solving and engaging students in reasoning has been demonstrated throughout this book in the student dialogues. Students' reasoning abilities were able to grow as students grappled with challenging problems in the different dialogue examples. In a problem-solving classroom, simply getting the answer is not the only important goal; rather, the thinking and reasoning necessary to problem solve are as important as the answer. Problem-solving activities provide a vehicle for you to learn how your students are thinking and reasoning about mathematics. Problem-solving tasks that are relevant to students help them see how mathematics connects to their real-world lives.

Teaching mathematics via problem solving can be likened to inquiry-based instruction. This type of instruction supports rich discussions that enable you to assess how your students reason. The traditional method of direct instruction in which the teacher presents rules and algorithms for students to memorize inhibits students' development of reasoning. Mathematics taught through problem solving provides students multiple opportunities to investigate the mathematics as they learn. These experiences uncover important ideas and allow students to think deeply about mathematics concepts and skills. When problem solving becomes the way you teach mathematics, it strongly supports the student discussions that allow you to assess your students daily.

Student Discussions

Informally observing students during student discussions is critical to determining their development of reasoning skills. The student dialogues in this book model ways for you to assess your students' reasoning skills during discussions. Following are examples of how teachers were able to assess students' ability to reason during their student discussions:

■ *Teachers asked students to explain their reasoning in arriving at a solution to a problem.* It is important to remember that students may have a correct solution, but their reasoning in reaching that solution may be flawed. Unless we question our students about their thinking in solving a problem, a misconception might not be revealed. Questioning our students provides us with information that can be used instructionally to redirect students' misconceptions toward reasoning that makes more sense. Make it a habit to ask all of your students to explain their reasoning, which provides you with important assessment information.

■ *Teachers asked students to share their strategies for solving a problem.* After a student shares a strategy and explains the reasoning for the strategy, encourage other students to share any strategies that may be different. Students will learn about a variety of ways to reason about problem solving, and you will find out about students' individual problem-solving strategies and reasoning skills. As students become familiar with a variety of strategies in problem-solving, they have a larger repertoire from which to draw when a particular strategy does not work. They can try another approach to a problem rather than give up.

■ *Teachers provided students time to work with a partner or in a small group.* When beginning your student discussions, you may find that some of your students are hesitant to speak in front of the whole class. However, in smaller groups students generally feel less intimidated and are more apt to share their reasoning. Use these opportunities to observe and listen to how your students are thinking and reasoning about a question or task. This is an opportune time to observe students who are not willing to participate in a whole-group discussion.

■ *Teachers often asked students to clarify their own thinking when statements were not clear.* Students should recognize that reasoning involves convincing another person about one's thinking. If reasoning is not clear, it is expected that students will be asked to explain in more detail. Students need to be asked for clarification when reasoning is flawed as well as when it is correct but unclear or difficult to understand. We want students to realize that discourse is a natural part of the classroom. Students will be asked for clarification and explanation when thinking is correct, not just when it is incorrect.

■ *Teachers often asked students to restate another student's reasoning.* This practice encourages students to listen carefully to their classmates' reasoning explanations. When asked to restate reasoning, a student must reflect on the reasoning shared in order to restate the thinking in her own words. If the original reasoning is correct but hard to understand, another student's rewording of the reasoning helps to clarify it. If a student's reasoning is flawed, you learn how the student is thinking about the mathematics. Encouraging students to revise flawed reasoning provides the student whose reasoning is flawed an opportunity to reflect and redirect his thinking and reasoning.

The reasoning activities on the CD, and the discussions that follow the completion of these activities, serve easily as informal assessments to determine the level of your students' reasoning abilities. These activities allow all students to demonstrate their knowledge regardless of where they are in their understandings of the mathematics being taught and how they reason about important mathematical ideas and relationships. The CD activities encourage students to focus on the thinking, reasoning, and processes that are essential for solving problems. (See Figure 6–1.)

During the discussions you facilitate with your students, it is unlikely that you can assess every student every day. However, before any discussion begins it is helpful to select a small group of students that you want to observe during that day's mathematics discussion. While all students are discussing the mathematics in the activity, make mental notes about the reasoning of the targeted students as well as their understanding of the mathematics in the lesson. These mental notes can then be recorded soon after the discussion as anecdotal notes (e.g., on address labels or index cards or a copy of a seating chart that has room for observation comments). Gather additional data from different targeted groups of students throughout the week.

Partner or Small-Group Work

If you are asking your students to work in small groups or with a partner, this is an excellent time to observe your students to assess their understanding of concepts and reasoning skills. Record observation notes as you listen to students discuss and work on a problem or task. As you facilitate the small groups, ask students probing questions to gather additional information about how they are reasoning as they complete the tasks. If you have students who are hesitant to share their thinking in a large-group setting, they may be more likely to share their reasoning in a small-group situation that is less intimidating. Then, when groups share their reasoning, students are sharing thinking of several students rather than placing their individual reasoning on the table for the large group to critique. This strategy might be particularly useful at the beginning of the year when you are still establishing the classroom norms for sharing and students are beginning to appreciate the reasoning contributions of their peers. As with the whole-group discussions, think about which students you want to collect data from during that day's mathematics lesson and then record this information discreetly. You may be recording data from only one small group during a lesson, but it is important to visit all the groups as they are problem solving.

Your observations of students as they work in small groups can highlight reasoning that you want to discuss in a whole-class environment. You can note different approaches that groups of students make and decide how to sequence those approaches in a class discussion in order to help students strengthen and enhance their reasoning abilities. Sequencing reasoning explanations from least sophisticated (perhaps just using examples) to more sophisticated (such as a deductive argument) can help students learn to compare and critique different justifications; over the course of the year, all students can grow in their ability to reason mathematically.

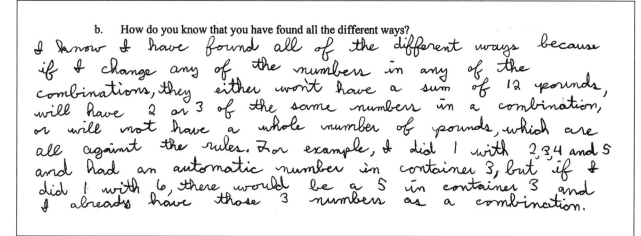

Figure 6–1 *A student's reasoning to explain why all the combinations for the* Containers of Nuts *problem have been found.*

CLASSROOM-TESTED TIP

When students are working individually or in small groups to complete problem-solving tasks, we are presented wonderful opportunities to observe them as they work and talk with each other. Following are suggestions for how to focus your observations when you are assessing students' reasoning. We recommend that you narrow your focus to only one or two of the following suggestions in a particular lesson:

- Does the student have a plan for thinking about and solving the task?

- Is the student's strategy an appropriate one for the task?

- If the student's strategy did not work, did the student apply a different strategy?

- Is the student able to self-monitor her thinking and reasoning?

- Does the student's representation accurately convey the reasoning used in thinking about the task?

- Is the student's explanation of reasoning clear and precise?

Written Responses

Students' written responses provide us with a valuable source of information about their development of reasoning. Written responses provide visual evidence of how students reason about the mathematical concepts and skills we teach. Your analysis of these responses helps you make decisions about your mathematics program and instruction. In addition, written responses are concrete evidence of students' learning that can be used as a way to communicate to parents during conferences about how their child is reasoning mathematically and about her understanding of the concepts and skills being taught.

Occasionally ask students to share their written responses with their classmates. Students' sharing of their problem-solving strategies and reasoning enables others to hear a variety of ways to think. When other strategies are described, encourage students to test the strategies on additional problems. This is similar to asking students to restate another student's reasoning during a discussion. Students will be discovering important ideas about different strategies that can lead to generalizations. The sharing of written responses provides another opportunity to assess students' reasoning skills (see Figure 6–2). Students can share written responses via transparencies (or through use of a document camera or Elmo) or on chart paper. One advantage of chart paper is that multiple strategies can be posted simultaneously so that students can compare and contrast them.

Students' representations are another form of informal assessment that allows us to understand more about the reasoning abilities of our students. Students' concrete, pictorial, and/or symbolic representations show us how they are organizing and communicating their thinking and reasoning. Observing how students utilize these representations of mathematical ideas and relationships provides us with valuable

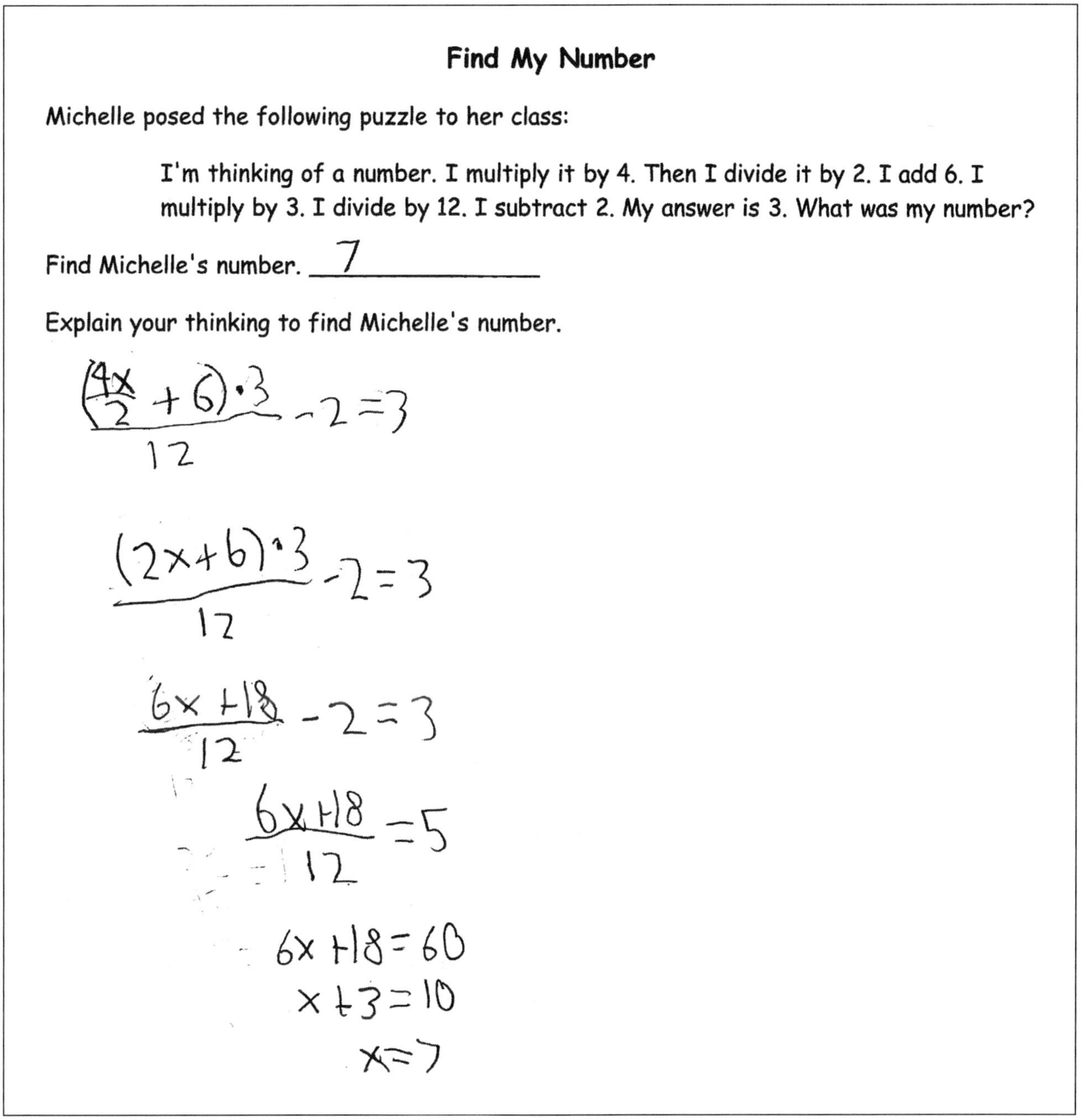

Find My Number

Michelle posed the following puzzle to her class:

> I'm thinking of a number. I multiply it by 4. Then I divide it by 2. I add 6. I multiply by 3. I divide by 12. I subtract 2. My answer is 3. What was my number?

Find Michelle's number. ___7___

Explain your thinking to find Michelle's number.

$$\frac{\left(\frac{4x}{2} + 6\right) \cdot 3}{12} - 2 = 3$$

$$\frac{(2x + 6) \cdot 3}{12} - 2 = 3$$

$$\frac{6x + 18}{12} - 2 = 3$$

$$\frac{6x + 18}{12} = 5$$

$$6x + 18 = 60$$

$$x + 3 = 10$$

$$x = 7$$

Figure 6–2 *A student represented her reasoning to the* Find My Number *task by expressing the problem symbolically with variables.*

information. In order for students to make sense of the mathematics, they will need to reason about the mathematics in the representation. Students' representations aid them in visualizing generalizations and conjectures they have made, such as a student who makes a graph, creates a table, and obtains a numerical solution (refer back to Figure 1–3).

Diagnostic Interviews

Diagnostic interviews are an effective method for gathering data about a student's ability to reason. A diagnostic interview is a one-on-one discussion with a student and should be short. A diagnostic interview helps you identify how a student is reasoning about a specific mathematics concept or skill as well as any strategies being used in the reasoning process. Many teachers find it manageable to schedule several interviews daily while other students are working independently on a task; over the course of a week or two, every student can have a few minutes of individual time with the teacher. Generally, five minutes is ample time to obtain the assessment information needed from a student. When you allow yourself time to administer a diagnostic interview with a student, you are more likely to identify the misconceptions or difficulties he may be experiencing with the mathematics he is learning. The reasoning students share during an interview helps to demonstrate the depth of their understanding about a concept or skill. In addition, the interview provides you with information that helps you plan instruction to meet the needs of all your students.

CLASSROOM-TESTED TIP

Before doing a diagnostic interview, plan the questions and materials ahead of time. Begin an interview with a quick warm-up, a task that the student is able to do with success. Then ask the student questions that will provide evidence of her ability to reason. Many of the questions you will want to ask are those provided in Chapters 1, 3, and 4. Also, the questions modeled throughout the student dialogues will be helpful to use during an interview assessment. A diagnostic interview is not a time to teach; rather, it is a time to observe and listen to what students know and are able to do, and how they are reasoning about important mathematical ideas and relationships. It is essential that you are neutral in your responses. At first this may be difficult for students to accept if they are used to a different type of response, such as *Great!* or *Super!* Smiles, frowns, or other responses will cause students to think that the answer they gave is right or wrong. And, of course, allowing wait time during any interview is extremely important.

Additional Ways to Assess Students' Reasoning

Another effective strategy to assess reasoning is to ask questions that are open-ended. These questions encourage students to reason more thoughtfully because they are not as tightly defined as traditional questions, which usually generate one right answer. Open-ended questions can be answered using a variety of methods or can have many right answers. Consider the *Containers of Nuts* problem in Chapter 1. In this problem, students were asked to list the different combinations of containers that would be possible if a person had 12 pounds of nuts and three containers so that no two containers could hold the same amount. Even though several constraints were attached to the problem, students were able to find a number of combinations. In students' reasoning explanations that stated how they knew all the possible combinations had been found, the teacher was able to assess how they were thinking and what strategies they were using to solve the task (refer back to Figure 6–1).

Open-ended questions in which you ask students to write a word problem for a computation can provide insight into the depth of students' conceptual understanding. For instance, Chappell and Thompson (1999a) report results when they asked middle grades students to write a word problem that could be solved by computing 5.68×2.34. They found that some students wrote a problem that required addition, others wrote a problem in which the context did not make sense for a decimal (e.g., 5.68 books), and others wrote appropriate real-world problems. Even though students could perform the computation numerically, some had difficulty determining situations when such a computation would be needed.

In another context, Chappell and Thompson (1999b) gave students the following task: *Draw a figure whose perimeter is 24 units.* They found that some students confused area and perimeter while others had misconceptions about the meaning of perimeter. The insights they gained about students' thinking might not have been evident had they just given students a figure with dimensions labeled and asked them to find the perimeter. By keeping this task open-ended, which allows many figures as possible answers, students are given an opportunity to demonstrate their own understanding and models about perimeter. Although the task does not specify a particular shape for the figure, many students drew rectangles, perhaps indicating how frequently students explore perimeter only through common figures. When too many students only consider common figures, the teacher can design a task to encourage students to broaden their set of possible solutions to include non-standard shapes.

Open-ended problems such as the two used by Chappell and Thompson are relatively easy to create. Instead of starting with the context and expecting students to find the answer, give students the answer and expect them to create the context. In essence, we are using the format of the popular television game show in which contestants provide a question to go with a specified answer.

The open-ended questions just discussed are essentially constructed-response questions that require students to apply what they know about a mathematics concept or skill. Other constructed-response questions ask students to explain why a solution makes sense or whether it is valid. For example, the first part of a constructed-response question may ask students to identify an appropriate unit of measure to determine the length of a classroom, and the second part of the question might ask students to use what they know about units of measure to explain why their solution makes sense. This information can be represented using words, numbers, or pictures. The answers to the last part of the question provide you with valuable information about students' understanding of the concept. Misconceptions that you observe your students making frequently can become the basis for a constructed-response question in which students analyze a solution to determine its validity. For instance, Figure 6–3 highlights a task from the CD in which algebra students consider a possible solution to a problem in which a common error is made. Other examples have been provided earlier in the book (see Figures 3–1, 3–2, 3–3, and 3–4).

Performance tasks are a beneficial form of assessment when your students are routinely investigating the mathematics they are learning. Performance tasks are generally open-ended and require more thinking and reasoning than a traditional assessment. They can be administered to pairs of students so you can observe them talking as they work on the task together. These tasks involve the manipulation of materials for students to make decisions and reason about the task. Students are actively participating in this type of assessment. Performance tasks may take students a considerable amount of time to complete, particularly if they need time to think about the problem, try several approaches, and then write about their reasoning. Such tasks might be used as problems of the week or as assessments that students complete at home over several days and then return. Extended performance tasks may be more difficult to use on a formal classroom assessment in middle school if students are expected to complete it as part of a typical classroom test. We use these problems because we want students to think and reason about them; we should not expect them to answer such problems quickly. Students should not be expected to complete performance tasks without sufficient time to think and reflect on their response.

Students in grades 6 through 8 are capable of assessing their own learning. This process of self-evaluating helps students assume responsibility for their own learning. They will be monitoring their progress in learning and reasoning about mathematics. Many of the student dialogues demonstrate ways that students can self-evaluate.

Equations with Products

Here is Megan's work on an algebra problem:

$n(n - 4) = 12$

$n = 12$ or $n - 4 = 12$

$n = 12$ or $n = 16$

a. Is Megan's work correct? Explain your reasoning.
   No, because $12(12-4) \neq 12$, and $16(16-4) \neq 12$

Figure 6–3 *A student's reasoning on the* Equations with Products *task, an assessment item in which students reason about a common algebra mistake.*

A simple rubric that encourages students to reflect on their reasoning is included on the CD.

Self-monitoring can be facilitated through the use of journals at different times throughout a unit of study (see Chapter 5). Near the end of a unit of study, ask students to list topics they feel they understand well, those on which they are shaky, and those that they still do not understand. Students can use their lists to guide the review and summary at the end of a unit before a summative assessment.

CLASSROOM-TESTED TIP

To help students develop self-monitoring skills, or the ability to think about their thinking, encourage them to ask themselves questions such as the following when they are investigating problems and tasks:

- What do I know about this problem or task?

- What strategy or strategies will help me?

- How can I connect this task to something I already know how to do?

- What do I think the answer will be, and why do I think this?

- Where should I begin?

- Will my strategy for solving this problem make sense to others?

- Do I need to revise my reasoning?

- Do I need to support my reasoning with more examples?

- Will the explanation of my reasoning be clear to others?

- Is my solution complete?

The closure of a lesson provides you with an additional opportunity to assess students' learning. This is the time for students to share what they have learned in that day's lesson, and our role is simply to be the facilitator of this brief discussion. Allowing students to summarize their learning in a student discussion provides you with another informal assessment of what they know. If students have any misconceptions, they are often revealed during the closure of a lesson. These errors in thinking can be noted and clarified so that students do not leave class with confused understandings, and these errors can become the basis for the beginning of the next day's lesson.

Exit cards provide a quick snapshot of students' reasoning and can be easily implemented into a lesson (Daniels and Zemelman 2004). They are given to students at the conclusion of a lesson; however, exit cards should not be confused with the closure that was just described. Exit cards differ from closure in that these cards are completed individually by students. They are easy to prepare by recording a question that requires students to reason about an idea or relationships investigated in that day's lesson.

The information we collect about our students' learning should be useful data that enables us to make sound instructional decisions for students. We must be able to interpret assessment information from a variety of sources so that we are able to measure how students reason in different ways. Integrating informal formative assessments into our daily teaching will help guide the instructional decisions we make for all students.

Development of Students' Reasoning Skills

The following are characteristics that students may demonstrate as they develop reasoning skills. These descriptors are meant to be general observations about students' progress in reasoning.

- Students with undeveloped reasoning abilities provide little or no evidence of a strategy or way of reasoning about a problem. If they do describe a strategy or explain their reasoning, it does not help them reach a solution that makes sense. They have difficulty understanding the reasoning of others.

- Students who are beginning to develop reasoning skills are implementing strategies and reasoning skills that are partially developed; however, the strategies or reasoning they are using do not support them in arriving at a fully developed solution.

- When students' reasoning becomes more developed, they begin to use a strategy and reasoning that guide them to a correct solution; however, they do not attempt to apply multiple strategies to arrive at the same solution. Students' reasoning about an idea or relationship is based on one or a few examples, although supporting the generalization is difficult.

- Students whose reasoning is stronger are comfortable applying several strategies to reach a solution, and they understand which strategies are more efficient than others for solving and reasoning about a problem. They are beginning to understand that a generalization, or conjecture, should be justified with multiple examples, and they are able to develop informal mathematical arguments to do this.

- Students who are efficient and more creative in their development of multiple strategies also have different ways to reason about mathematical ideas and relationships. They are able to self-evaluate their reasoning for the purpose of refining it, and they evaluate others' reasoning in an attempt to make sense of the mathematics they are learning. Students are able to make conjectures and develop mathematical arguments about more complex ideas.

[These descriptors were adapted from the work of Bena Kallick and Ross Brewer in *How to Assess Problem-Solving Skills in Math* (1997, 125).]

95

*Assessing Students'
Reasoning*

Assessing Students' Reasoning with Rubrics

A rubric is an assessment tool often used by teachers on constructed-response items. Middle grades mathematics teachers have long assigned partial credit to students' responses on problems. However, a partial-credit scheme is not the same as a rubric. Partial credit is often assigned on an item-by-item basis. How many points a response receives can vary from item to item, so it is often difficult to compare students' performance across items. In contrast, a rubric is developed conceptually, with a given set of criteria used to assess performance at a particular level. If teachers use a consistent rubric, then they can compare students' reasoning abilities across multiple problems in multiple contexts (Thompson and Senk 1998). For instance, suppose teachers use a rubric with a scale from 0 to 4. One student has 2, 2, 2, 2, and 2 as the scores on five problems; another student has scores of 4, 2, 4, 0, and 0. Even though both students have a total of 10 points, we would argue that their knowledge and understanding are likely different. The first student makes some headway on every problem; in contrast, the second student generally solves the problem completely or does not make much progress.

Many states now have constructed-response items scored with a rubric as part of their state assessment program. If you teach in such a state, we suggest that you use the state rubric as your rubric throughout your instruction. This helps students become accustomed to the levels of explanation needed for different score levels.

It should be noted that the use of rubrics only makes sense when more than a single answer is expected or possible, or an explanation of thinking is required. Following are two general rubrics that can be used for assessing reasoning and constructed-response tasks. In the first rubric, there are three score levels. In the second, scores range from 0 to 4, with 3 and 4 considered successful responses and 0, 1, and 2 considered unsuccessful. Notice that clerical versus conceptual errors provide the distinction between successful and unsuccessful responses. A clerical error might be a minor error in notation, language, form, or computation so that the student does not obtain a correct result; nevertheless, the response clearly indicates the student understands the concept. Clerical errors are often careless errors, not errors in understanding.

Sample Rubric 1

3   Student states reasoning that is clear and makes sense to others. The reasoning is well developed with multiple examples of justification.

2   Student states reasoning that makes sense; however, it is supported by only a few examples of justification. Or the student may have a flaw in the reasoning but upon reflecting on the reasoning is able to revise the original thinking.

1   Student is unable to explain why a solution is correct. Or the student has a flaw in reasoning, and in reflecting upon it, is unable to revise the original thinking.

Sample Rubric 2

Successful Responses

4 Solution is complete and response explains reasoning thoroughly. The response might serve as a model response.

3 Student provides a reasonable solution or explanation, but minor errors occur in notation, form, or vocabulary. Any errors are considered clerical, rather than conceptual, in nature.

Unsuccessful Responses

2 Student provides a response in the proper direction that displays some substance or a chain of reasoning. However, the student may stop about halfway through completion of the problem. Errors are conceptual, rather than merely clerical, in nature.

1 The student provides a response that indicates some initial progress into the problem, but the student reaches an early impasse and displays minimal reasoning.

0 The student makes no mathematical progress on the problem. Any work is mathematically meaningless in relation to the problem.

Adapted from Thompson and Senk (1993).

At times, teachers may want to distinguish between students who begin a problem but make little or no meaningful progress and students who do not even start a problem. Willingness to at least tackle an unusual problem is part of mathematical perseverance that we want our students to develop because it reflects their mathematical disposition and attitude. In such instances, we use a score of 0 to indicate a mathematical attempt that is meaningless and some arbitrary symbol to indicate no attempt at all on the problem.

One caution should be noted about the use of rubrics. Assignment of grades, an issue that teachers in grades 6 through 8 face, should be separate from the use of rubrics. For example, in the second rubric, a score of 3 should not be considered 75%. After all, it reflects a successful response, so a grade of C or C– would not be appropriate.

In Chapter 1, we discussed the importance of explaining to students the expectations you have for them in communicating their reasoning. Just as it is important for students to understand why they will be expected to discuss, write, or represent their reasoning, it is also essential that they understand how you will assess their development of reasoning throughout the school year. A general rubric may be sufficient initially for your assessment needs and certainly easier for students to understand. However, you may later decide that a more specific rubric is necessary to assess your students' progress in learning how to make conjectures and develop mathematical arguments. These informal assessments of students' reasoning skills can guide you in how to best meet your students' needs. Assessing students' progress in their ability to describe generalizations about mathematical ideas, state conjectures, and make mathematical arguments will help you plan next steps to provide additional experiences to support their reasoning skills. These experiences build the foundational skills necessary for students' more formal reasoning in later school years. When we ask our students to reflect on their reasoning, it not only assists us in understanding how they are thinking but also allows them to self-evaluate their reasoning and evaluate the reasoning of their classmates.

Final Thoughts

Students in grades 6 through 8 are capable of reasoning at a higher level than their reasoning level in the elementary grades. It is essential that we understand how students reason in the middle grades so that we can plan instruction that enhances their reasoning abilities. Assessment must then become an integral and ongoing part of our mathematics teaching so that we are able to monitor our students' growth in reasoning to make instructional decisions that will affect all our students in positive ways.

The most efficient assessments are often those that can be done easily and informally. The discussions you have with your students will provide you with daily opportunities to monitor their development as mathematical thinkers. Students' written responses and representations will give you visible evidence of how they are reasoning that you can then share with their families. These multiple types of assessment enable us to understand more about our students' thought processes and the depth of understanding they have about mathematics concepts and skills. And we cannot overlook the importance of our students learning to self-monitor their reasoning as well as evaluate the reasoning of others. When students are provided these experiences, they will take ownership of their learning, which will support them in becoming powerful mathematical thinkers.

Questions for Discussion

1. The importance of assessing students' reasoning skills was the focus of this chapter. Many ways to do this were discussed. Which of these recommendations will you feel the most comfortable using? Why do you feel this way?

2. Which method of assessment are you the most uncomfortable using? Reflect on this yourself, or discuss with your colleagues possible ways to make this more manageable for you and your students.

3. Suppose you have observed that several of your students are experiencing difficulties in making conjectures that are clearly developed and understood. What types of activities would you plan to help these students strengthen this reasoning skill?

4. Locate information about your state's assessment program, if one exists. Are performance tasks part of the assessment? If so, identify any rubrics used to evaluate responses to those tasks.

7

Reasoning and Proof Across the Content Standards

Effective teaching conveys a belief that each student can and is expected to understand mathematics and that each will be supported in his or her efforts to accomplish this goal.

—National Council of Teachers of Mathematics,
Principles and Standards for School Mathematics

We have been focusing on how to help students develop the process skill of reasoning and proof and on how students' ability to reason supports their ability to make sense of mathematics concepts and skills. The interconnectedness of reasoning and proof with all the process standards is critical to support students' reasoning skills. Students communicate their reasoning in student discussions, in written responses, and through representations. Students' ability to communicate reasoning in a variety of ways enables them to become flexible in their mathematical abilities as well as reflective thinkers. As they reflect on their reasoning, and the understandings they are developing about mathematics, their reasoning and conceptual understandings are refined and internalized.

It is beneficial for students to represent their reasoning concretely, pictorially, and symbolically. These representations assist them to understand their thinking and reasoning about a mathematical idea or relationship, and they become tools that reveal misconceptions about a student's reasoning when they occur. The connections that students make among mathematical ideas support them as they develop deep understandings about those ideas. Not only do students strengthen their reasoning skills as they make connections among mathematical ideas, but when they reason about mathematical ideas connected to real-world situations, their reasoning is applied in practical ways that prepare them for the future. In this chapter, we use student dialogues to discuss how the reasoning and proof process standard interconnects with the other process standards as well as the content standards.

In Chapter 5, we discussed how the process standards overlap, and this is also true about the connections between the process standards and the content standards. Mathematics is often considered to be an abstract discipline, but students' application of the process standard of reasoning and proof allows them to make sense of mathematics. As students solve problems that are interesting and challenging, they naturally use intuitive, inductive, and deductive reasoning skills to think about a variety of problem-solving situations. Teaching mathematics through problem solving provides our students multiple opportunities to reason, create conjectures, and make mathematical arguments. For students to make this reasoning explicit, they must use the process standards of communication and representation. Students' ability to connect the mathematics they are learning to other mathematical ideas, across all disciplines, and in their everyday lives allows them to reason in a variety of settings that demonstrate application of mathematical ideas.

The National Council of Teachers of Mathematics (2000) has outlined content standards for students in grades 6 through 8 and has organized those standards in five content areas: number and operations, algebra, measurement, geometry, and data analysis and probability. Although in our teaching we focus on helping students develop critical understandings of this content, we must also support our students to develop their process skills of reasoning and proof, problem solving, communication, representation, and connections. This chapter provides examples of student dialogue and activities that illustrate the integration of content and process. We discuss how students reason about the various mathematical ideas and relationships presented. The activities discussed in the dialogues are aligned with the NCTM Standards and expectations for students in the middle grades (NCTM 2000). Additional resources to support you in implementing these activities are available on the CD.

Number and Operations

Elementary students in grades 3 through 5 work on number sense with a focus on whole numbers in the hundreds and thousands, multiplication and division, fractions, decimals, percents, and integers. The goal in the intermediate grades is to help students become flexible thinkers about all types of numbers and to nurture their number sense.

In the middle grades, students continue to work on developing number sense, with a particular goal to enhance and strengthen their understanding of rational numbers in all forms, including fractions, decimals, and percents larger than 100 and smaller than 1. Proportional reasoning is an essential part of the middle grades mathematics curriculum; in addition to its use with ratios and proportions, proportional reasoning is fundamental to work with linear functions (e.g., functions of the form $y = mx$, $m \neq 0$), with geometric concepts of similarity, and with data analysis when comparing samples of different sizes.

Throughout the earlier chapters, we have discussed how reasoning can be integrated with number sense activities. For example, in Chapter 5 (Figures 5–3 and 5–4) we illustrated students' work in finding two numbers with a sum of 75 and a product of 1400. Such a task requires students to think about how numbers are related. As students selected a pair of numbers whose sum was 75, they checked their product. Each product other than 1400 provided insight that enabled students to adjust their numbers until the appropriate pair was determined. In Chapter 3, we shared students' work when reasoning about the distributive property. In one case, students reasoned abstractly about a conjecture related to multiplying $3(x - y)$; and in another, they used a retail sales context to reason that finding a price after 30% off was equivalent to multiplying the original price by 70%. In the *Strawberry Smoothie* task in Chapter 4 (see Figures 4–3 and 4–4), students used proportional reasoning to determine which of two smoothie mixtures had a stronger strawberry taste and then to determine the amount of strawberries needed to make a larger quantity of smoothie with the same taste. Each example illustrates that reasoning can be a natural part of mathematics instruction throughout the number and operations content strand. Students' use of number sense to think about mathematics supports their ability to reason about the ideas and relationships between and among numbers.

The importance of reasoning and problem solving in developing students' deep understandings of mathematics has been stressed throughout the book. These understandings are further developed when learning is supported by the remaining process standards of communication, representation, and connections as discussed in Chapter 5. In this chapter, we provide a student dialogue for each of the content standards to illustrate how reasoning and proof enhances the ideas and relationships in that content. As you read through the dialogues, notice how the other process standards and/or content standards are integrated in a lesson's activity.

Lesson: *Factors and Multiples*

Students in grades 6 through 8 should be expected to use factors, multiples, and relatively prime numbers (numbers with a greatest common factor of 1) to solve problems. In the following *Going on a Field Trip* task, students investigate a problem set in a real context and then look for patterns in the numbers to explain the results they obtain.

> Ms. Roberts' science club is going on a field trip. There are 72 students in the club. Parents are helping to transport students in a combination of vans and cars. Ms. Roberts wants to be efficient, so every possible seat must be filled by a student. Find all the possible combinations of cars and vans that could be used to transport the students on the field trip in each case.
>
> 1. A car can carry 4 students and a van can carry 8 students.
> 2. A car can carry 4 students and a van can carry 12 students.
> 3. A car can carry 4 students and a van can carry 9 students.
> 4. Compare your answers in problems 1–3. How are they alike and how are they different? Why do you think any differences occurred?

TEACHER: (*The problem is displayed on the overhead for the entire class to see.*) I want you to take a few minutes and work with your neighbor on the problem in the first case—when a car carries 4 students and a van carries 8. (*Students are*

STUDENT: The first thing I did was see if 4 would go into 72. It divides it and I get 18. So you could have 18 cars. Then I found that 8 divides into 72, and so there are 9 vans. So one combination is 18 cars and 9 vans.

STUDENT: You can't have 18 cars and 9 vans at the same time because that would be too many people. Eighteen cars hold 18 times 4 or 72 people and 9 vans hold 9 times 8 or 72 people. But that's too many people. If you have 18 cars then you don't have any vans, and if you have 9 vans you don't have any cars.

TEACHER: We have two different points of view on the table for discussion. Someone else want to give their view?

STUDENT: If we make a T table, then 18 cars and 0 vans is one row and 9 vans and 0 cars is another row. I agree that you can't have 18 cars and 9 vans at the same time, or you'd have too many people.

TEACHER: *(addressing the first student)* What do you think about this reasoning? What do you think about putting your amounts on different rows? *(Student agrees that the two amounts should be separate entries in a table containing the combinations.)* What other combinations of cars and vans did you find?

STUDENT: We had 9 vans and we took away 1 van. So then we had 8 vans that hold 8 times 8 people. When we subtracted the 64 from 72 we had 8 people left, so we needed 2 cars. So another way is 8 vans and 2 cars.

STUDENT: We tried 6 vans and got 48 people. Seventy-two minus 48 is 24, so we need 6 cars. *(Students continued to discuss combinations randomly and record them in a table as shown in Figure 7–1.)*

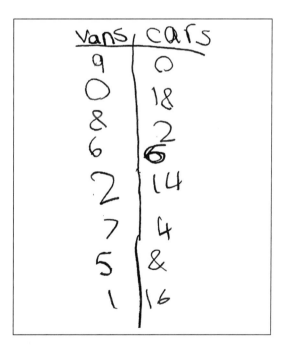

Figure 7–1 *Initial student work on the first part of* Going on a Field Trip, *recording combinations in a random fashion.*

TEACHER: Okay, we've got several combinations here. Do we have all of them? How can we tell?

STUDENT: We found some that aren't in the table so far. We remembered those problems where we had to make a list, and so we started with 0 vans and 18 cars and then went up by 1 van each time. Here's what we got (see Figure 7–2).

TEACHER: I like the way you've organized your work. The first table we have gives us a number of combinations. But remember the *Containers of Nuts* problem (see Chapter 1). One way we knew we had all the combinations was to organize our work systematically. So, while the first table is a great way to get us started in thinking about the problem to understand what is being asked, the second table helps us develop an argument that we have all the possible combinations. Let's see if you can use what we've done so far to find the possibilities for parts 2 and 3 in the problem. After you have a few minutes to work, I want to have someone record their solutions on the board and explain to us how you know you have all the combinations. (*Students are given a few minutes to work and then the students in Figure 7–3 present their solution to the second problem on the board.*)

STUDENTS: (*from Figure 7–3*) We started with 6 vans because 6 times 12 is 72. Then we started going down by 1 van each time. We found the number of cars went up by 3 each time. We think this makes sense because 3 cars is the same as 1 van.

TEACHER: I'd like you to explain a bit more what you mean by "3 cars is the same as 1 van."

STUDENTS: Well, 3 cars hold 12 people and 1 van holds 12 people. So 3 cars hold the same amount as 1 van.

TEACHER: Thank you for clarifying that 3 cars and 1 van hold the same number of people. What other combinations do you need to add to your table?

STUDENTS: We still need 1 van and 15 cars and then 0 vans and 18 cars.

TEACHER: Do the rest of you agree with their strategy? (*Lots of nods.*) What happened on the third problem?

STUDENTS: We started with 8 vans because 8 times 9 is 72. When we tried 7 vans, that took 63 people and there were 9 people left. You can't use cars because every seat won't be filled. Then we tried 6 vans and that doesn't work either because we had 6 times 9 to get 54 people, and then that leaves 18. The closest you can get with cars is 4 because 4 times 4 is 16. The first one we could get to work was 4 vans and 9 cars and then 0 vans and 18 cars.

TEACHER: Why do you think problem 3 was different?

STUDENTS: In the first two problems, the numbers work out even but in the third problem they aren't even.

TEACHER: I'm not sure I understand what you mean. Can you clarify that?

STUDENTS: Well, in the first problem 4 divides into 8. So 1 van is the same as 2 cars. In the second problem, 4 divides into 12 so 1 van is the same as 3 cars. But in the third problem, 4 doesn't divide into 9 evenly, so you can't make cars and vans equal. You have to go down 4 vans and then up 9 cars.

TEACHER: I think you've hit on an important relationship here. What do you know about 4 and 8 and 4 and 12 that is not true for 4 and 9?

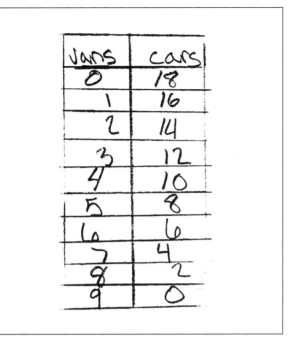

Figure 7–2 *Reworking of the first part of* Going on
a Field Trip *to show all the possible combinations.*

Figure 7–3 *Two students presenting their solution
to the second part of the* Going on a Field Trip *task.*

STUDENTS: Four is a factor of 8 and 12, so 4 divides into those two numbers evenly. Four isn't a factor of 9, so it doesn't divide into it evenly.

TEACHER: Okay, that's one way to think about it. If we say that 4 is a factor of 8 and 12, what's another way we can say that? We want to be able to think about these relationships in more than one way.

STUDENTS: Eight and 12 are both multiples of 4, but 9 isn't.

TEACHER: Okay, we've hit on an interesting distinction between these problems. I'd like you to take a few minutes and individually write about how these problems are alike and how they are different and why you think differences occurred. (See Figure 7–4 for a sample response.)

In this problem, students had an opportunity to work on problems in which factors and multiples determined the combinations; because 4 and 9 are relatively prime (i.e., 1 is their only common factor), students had fewer combinations for the third

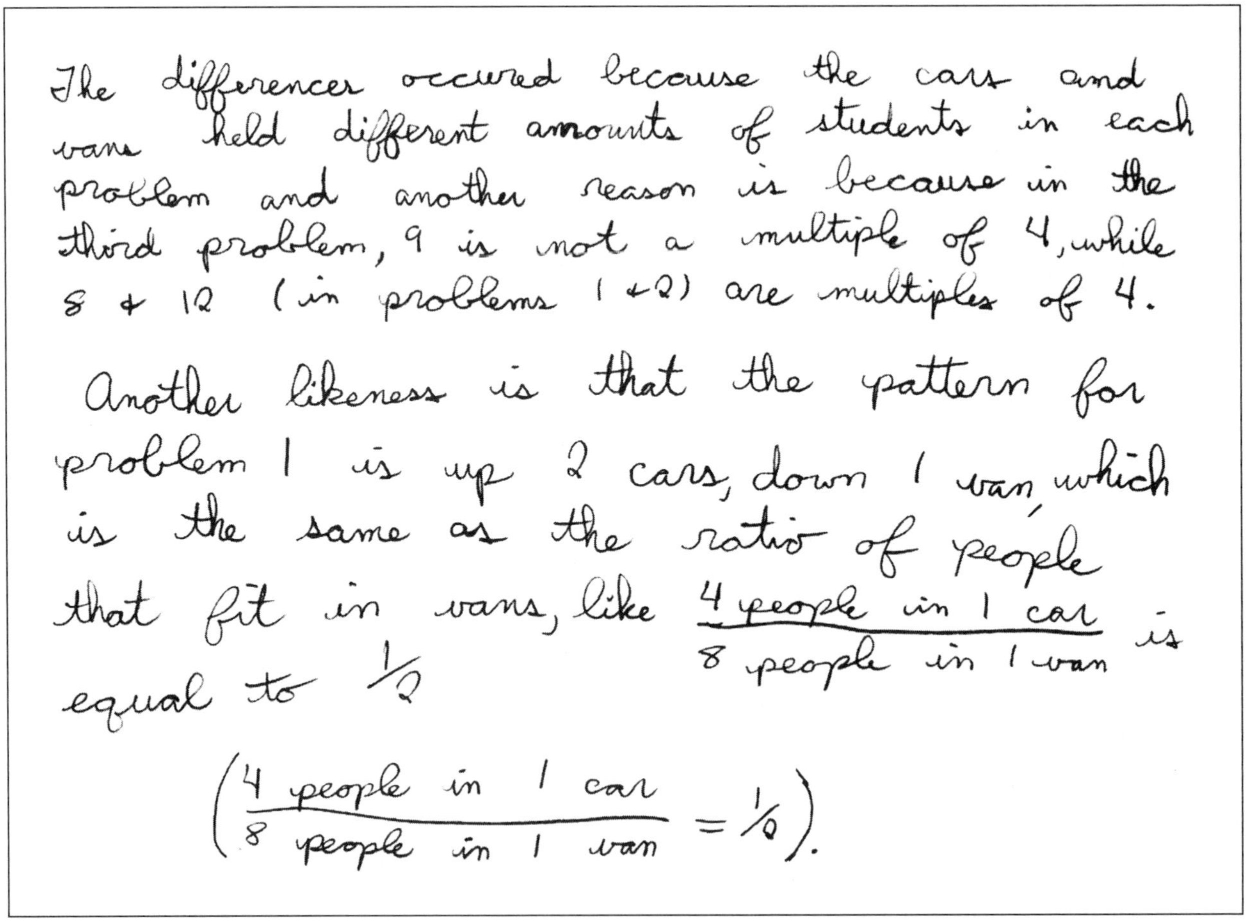

The differences occured because the cars and vans held different amounts of students in each problem and another reason is because in the third problem, 9 is not a multiple of 4, while 8 + 12 (in problems 1 + 2) are multiples of 4.

Another likeness is that the pattern for problem 1 is up 2 cars, down 1 van, which is the same as the ratio of people that fit in vans, like $\frac{4 \text{ people in } 1 \text{ car}}{8 \text{ people in } 1 \text{ van}}$ is equal to $\frac{1}{2}$

$$\left(\frac{4 \text{ people in } 1 \text{ car}}{8 \text{ people in } 1 \text{ van}} = \frac{1}{2} \right).$$

Figure 7–4 *Student reasoning about how the problems in* Going on a Field Trip *are alike and different.*

problem. When students wrote about their comparisons, they not only commented on concepts related to multiples, but they also commented on the ratios evident in the problem.

Notice that the student's language in describing the pattern observed in the tables (up 2 cars, down 1 van) mimics language related to slope. Students could have written equations for each problem and graphed the results. For instance, if c represents the number of cars and v represents the number of vans, then the three problems correspond to the equations $4c + 8v = 72$, $4c + 12v = 72$, and $4c + 9v = 72$, respectively. The graph of each equation is a line with a negative slope. If the horizontal axis is c and the vertical is v, then the slope in each case is $-\frac{4}{8}, -\frac{4}{12}$, and $-\frac{4}{9}$, respectively; these correspond to the patterns of change between vans and cars observed in each case.

As teachers, we need to look for connections between different areas of mathematics. Even if students are not studying algebra, they might graph the relationships in their three tables for cars and vans and compare the shapes of the graphs. We are helping to lay foundations for later formal study of these same concepts.

As evident in the dialogue and in examples throughout this text, reasoning occurs naturally throughout the number and operations content standard. The CD contains several activities that focus on different ways to engage students in reasoning about number. These activities are not meant to be exhaustive of the types of activities possible in this strand, but are meant to spark ideas from which you can build further activities depending on the experiences of your students. As we have emphasized previously, the discussions that occur during or following the completion of any activity help solidify the understandings students are developing. When students make their reasoning about number and operations visible in daily student discussions, in written responses, or in representations, they are strengthening their reasoning skills.

Algebra

NCTM recommends that the content of algebra be taught in all grades, prekindergarten through grade 12. We must provide our students with many opportunities to reason algebraically throughout the school year. "Rather than the algebra you may remember from your high school days, the algebra intended for K–8 focuses on patterns, relationships and functions, and the use of various representations—symbolic, numeric, and graphic—to help make sense of all sorts of mathematical situations" (Van de Walle and Lovin 2006, 290). An important focus for students in grades 6 through 8 is the development of their ability to use variables to generalize patterns as well as represent equations. In addition, students in grades 6 through 8 should be able to find equivalent forms of algebraic expressions and model situations with multiple representations, including tables, graphs, and symbols.

Students are using algebraic reasoning as they examine repeating and growing patterns, both numeric and geometric patterns. In the *Build That Figure* task in Chapter 2, students extended a pattern, both to terms easy to build and to terms further in the sequence, and provided different ways to generalize the pattern (see Figure 2–3). In the process, students generated multiple expressions that were equivalent to each other, despite the difference in their appearance. In the *Mystery Number Riddle* task, students used variables to develop a symbolic argument to show that the final result is always twice the original number. Some students used boxes for the original mystery number and dots for the numbers added or subtracted (See Figure 2–5); others used variables to represent the same argument (see Figure 5–6). The different representations reflect the fact that middle grades students are at different comfort levels with the use of variables depending on their prior experiences. Even students who may use variables in some situations, such as to generalize a pattern, may be less comfortable using variables in an argument (as in Figure 2–5). Throughout the middle grades, we want to provide opportunities for our students to reason about mathematics as they develop fluency with symbolic representations. As more students take a formal course in first-year algebra by the end of eighth grade, it is important that foundational activities be provided early in the middle grades years to ease the transition to formal algebraic study.

Students in grades 6 through 8 are able to look for and apply relationships among numbers as they determine whether conjectures are true or not. In Chapter 3 (see Figures 3–1 and 3–2), students investigated a conjecture about a possible commutative property of division. Although they found some cases where the conjecture was true, counterexamples proved that the conjecture was not true in general. Students realized that examples alone cannot prove a conjecture true in all cases, but a single counterexample can disprove a conjecture.

In Chapter 6 (see Figure 6–3), students reasoned about a common misconception for obtaining a solution to an equation. As students learn to compute and operate with symbols, there are many misconceptions that appear when students have fragile understandings. These misconceptions provide opportunities for us to engage students in reasoning about algebra. Through talking with their peers and comparing solutions, students can identify valid solution strategies as well as those that are invalid. As students reason about what is invalid about a particular approach, they will become less likely to have those misconceptions themselves.

By providing multiple experiences in which our students are engaged in reasoning about a variety of algebraic concepts, we are helping them develop the foundational ideas needed to understand formal algebra. To help students develop additional algebraic reasoning skills, provide experiences that enable them to

- reason about balance in equations;

- solve problems in a variety of ways;

- represent mathematical ideas and relationships in a variety of ways, including symbolically;

- reason about functional relationships, including proportional relationships;

- use inductive and deductive thinking and reasoning; and

work with variables that have multiple values, variables that represent a relation-ship between two sets of values, and variables that represent unknowns.

Lesson: *Solving Equations with Tables*

The following lesson, based on the task *Fund Raising with a School Play*, provides opportunities for students to reason with equations as well as tables in a real-world context.

> The drama club at South Artsville Middle School raises money by sponsoring a school play every year. On Friday evening, the performance hall costs $200 and tickets sell for $4. For the matinee on Saturday afternoon, the performance hall costs $75 and tickets sell for $3. How many tickets must be sold so that the Friday evening performance and Saturday matinee raise the same amount of money?

TEACHER: Look at the problem I have up on the board. I'd like someone to read the problem out loud. (*She calls on one of the students who volunteer to read.*) Before we focus on the problem, I'd like to ask you a different question: "How many tickets must be sold for each performance so the drama club breaks even? How do you know?" Anyone have an idea on how we might answer this question?

STUDENT: What does "break even" mean?

TEACHER: Can someone help us out? What do you think it would mean to break even?

STUDENT: We don't make a profit and we don't have a loss.

TEACHER: Okay. So what does that mean in the context of the problem? How would you determine when you don't have either a profit or a loss?

STUDENT: The amount you make for tickets has to be the same amount as what you spend for the hall. On Friday, I know that 50 times 4 minus 200 is zero, so that means 50 tickets is my break-even point. On Saturday, I have to sell 25 tickets because 25 times 3 minus 75 is 0.

TEACHER: So, you found the value where the total amount earned from the tickets is the same as the cost of the hall. How did you come up with 50?

STUDENT: I know the cost of the tickets is 4 so I divided 200 dollars by 4 and got 50. Then 50 times 4 minus 200 is 0, and I have no profit or loss.

TEACHER: Okay, so we need to find where the amount of money collected from tickets equals the cost of the performance hall. How could you use your explanation to write an equation that would tell us the amount of profit or loss for any number of tickets that we might sell? (*Teacher waits a few seconds for all students to think about her question in relation to the student's response.*)

STUDENT: I would take the 4-dollar cost of the ticket and multiply by the number of tickets sold and then subtract the cost of the hall. So if n is my number of tickets, I get $4n - 200$ for the amount the club makes.

TEACHER: How would you tell if you have a profit or a loss?

STUDENT: If it's positive, that's profit. Negative means a loss.

TEACHER: Can someone else use the same idea to explain how to find the profit or loss on Saturday afternoon for any number of tickets we might sell?

STUDENT: I would just do 3 times *n* minus 75. (e.g., 3*n* – 75)

TEACHER: Now that we have expressions for the amount we make for each performance, I'd like you to complete the table I've put on the board so we can find the amount of money the drama club raises from each performance. With your partner, I want you to fill out the first four rows. Then I want you to think about how using the table can help find the numbers of tickets we would need to sell for the two performances to raise the same amount of money. (*Students are given a few minutes to work together.*) I'd like someone to record their work on the board for all of us to see so that it can be compared to our own work. (*See Figure 7–5.*)

STUDENT: I just continued my table till I had 125 tickets, and then they both raised 300 dollars.

TEACHER: You have some extra columns in your table. What do those columns mean?

STUDENT: This column (*pointing to 0, 100, 200, 300, 400, 500*) is the amount collected from selling the tickets. Then this other column (*pointing to –200, –100, –0, +100, +200, +300*) is the amount the club raised.

TEACHER: Okay, so you just put in columns to show the total amount of the tickets before you subtracted the cost of the hall. Is that right? (*Student nods.*) And tell us again how you used the table to decide when the two performances raised the same amount.

STUDENT: I wanted the amounts to be the same in the two columns. So when they both had 300, the two raised the same amount. This happened when they sold 125 tickets each. (*The teacher had accomplished her goals for the lesson this day. Students had made sense of a real problem and learned new vocabulary of break even in that context. They had extended a table and used it to determine when two expressions were equal. They had also expressed the situations symbolically. But the teacher also knew that discussion about the problem still had much to offer in helping students reason about a situation and make connections among symbolic, graphical, and tabular representations. So, she decided to revisit this problem on the next day.*)

Number of Tickets Sold	Amount Raised on Friday		Amount Raised on Saturday	
0	0	0	–200	0	–75
25	100	–100	75	–0
50	200	–0	150	+75
75	300	+100	225	+150
100	400	+200	300	+225
125	500	+300	375	+300

Figure 7–5 *One student's table to show the amount raised from selling different numbers of tickets.*

TEACHER: I want us to revisit the *Fund Raising with a School Play* problem from yesterday and look at that problem using our graphing calculators. Yesterday, we found that we could write two equations to represent the amounts the club raised on each night. For Friday night, Profit = $4n - 200$ and for Saturday afternoon we had Profit = $3n - 75$. If you want to know when the two performances raise the same amount of money, what could you do with these two equations?

STUDENT: We could make them equal to each other like this. (*Student then writes on the board $4n - 200 = 3n - 75$ so $1n = 125$.*)

TEACHER: That's exactly what we want to do. In this case, the equations had relatively nice numbers. Sometimes the numbers are a bit messier, and so we might want to use technology to help us. Let's put both equations into our graphing calculator using **Y =**. (*Students put the equations into their calculator while the teacher puts them into the calculator hooked into her overhead panel.*) We haven't used the TABLE feature of the calculator before, but the calculator can do what you did by hand with your table. Let's see how we can do that. (*Teacher gave some specific directions for the calculator model she had in the classroom, a TI84.*) Yesterday, we wanted to find where the two columns had the same amount. So we just need to scroll down the table until we find where the two columns for Y have the same value. Did you get the same number of tickets we found yesterday? (*Students had some time to look at their table of values.*) (See Figure 7–6.) How could you use the table to find when the club earns $2000 from the Friday night performance?

STUDENT: We just go down the table till we find 2000. That happens when x is 550 tickets.

STUDENT: I like these tables. This is a lot easier than trying to solve those equations.

TEACHER: Why do you think that?

STUDENT: Cause I sometimes make mistakes with my positives and negatives.

TEACHER: So the calculator would give you one way to verify that the solution you obtain symbolically is correct. Let's graph the two equations. How would you use the graph to find out when the two performances raise the same amount of money?

STUDENT: We just have to find when the two graphs cross. We need to know the coordinates of the intersection point. I got $x = 125$ and $y = 300$. How do I know which one is the solution? (See Figure 7–6.)

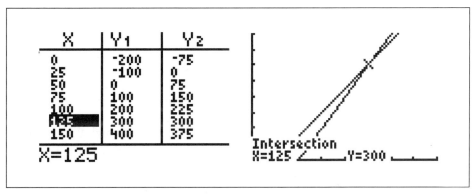

Figure 7–6 *Table and graph (window $0 \leq x \leq 200$, $0 \leq y \leq 400$) to explore the* Fund Raising with a School Play *problem.*

TEACHER: That's a great question. Anybody have an idea? (*No response.*) Think about what your variables mean.

STUDENT: Oh, I get it. The x is the number of tickets and the y is the total amount we earned. So they have the same value for 125 tickets. That's what we got before.

TEACHER: So, we now have three different ways to solve equations: by solving our symbolic equation, by looking at values in a table, and by graphing. Even if you have difficulty in solving an equation by one of these methods, you might be able to solve it using one of the other methods. You want to think about which method makes the most sense for a particular problem.

During this lesson, students were investigating, describing, representing, and explaining the relationship of equality as they determined when two expressions had the same value. When students solved this problem symbolically ($4n - 200 = 3n - 75$), they worked with a single linear equation. However, when they transitioned to a graphing calculator, they worked with a system of equations ($y_1 = 4x - 200, y_2 = 3x - 75$). As students use such technology, they will often make these connections between solving a single equation symbolically and solving a system with technology. Technology has the potential to help many students succeed in understanding and reasoning with algebraic concepts. Some of our students are likely to struggle initially as they develop symbolic procedures to solve equations. But technology provides other avenues to help students solve equations and to reason about important connections related to the meaning of a solution. Do students who can solve a symbolic equation but cannot understand what that solution means via a table or a graph really understand the meaning of a solution? Our students need to be able to solve equations through various approaches to have the robust understanding we desire.

Graphing calculator technology provides many opportunities for students to reason algebraically. Students could graph multiple equations with different costs for the tickets but the same cost for the hall to reason that the cost of the ticket influences the slope of the graph; the higher the cost, the steeper the graph. Students could also graph multiple equations with the same ticket cost but different costs for the performance hall to reason that the cost of the hall influences the y-intercept of the graph and that the lines for all equations with the same ticket cost are parallel. Whether students investigate such relationships via graphing technology or via paper-and-pencil graphs, such extensions of the *Fund Raising with a School Play* task provide additional opportunities for students to reason about connections among symbolic and graphical representations.

Geometry

The reasoning skills that students are developing in grades 6 through 8 serve an important purpose in the learning of geometry. These reasoning skills support students in exploring and thinking about geometric situations that are now more complex than in the elementary grades. It is appropriate for students to continue investigating ideas about geometry in which they identify, compare, classify, and analyze both two- and three-dimensional shapes, including an analysis of the properties of figures.

The van Hiele model of geometric thought (Crowley 1987) describes five levels of geometric understanding through which students progress as they reason about geometry. At the lowest level, visualization, students view figures in their entirety and base decisions on how figures look. At the second level, analysis, students begin to consider the characteristics and attributes of shapes (e.g., in all parallelograms the opposite sides are congruent). At the third level, informal deduction, students begin to reason about class inclusion and make informal arguments about properties of figures (e.g., all squares are rectangles). The goal of teachers in the middle grades should be to help students progress through these levels so that they are prepared for a formal geometry course in high school, in which proof is often a major focus and students are expected to operate at the deduction level in the van Hiele model. Research indicates that student success in such a geometry course depends on the level of geometry knowledge entering the course (Senk 1985). So, our goal relative to geometry is to prepare students for the deductive reasoning that has been a hallmark of such courses for many years. Our focus on reasoning about geometry in the middle grades can help students build that bridge from informal geometry in elementary school to more formal geometry in high school.

Students in grades 6 through 8 are likely to need hands-on experiences with geometric figures as much as possible, in addition to work with visual representations. Definitions are an important part of the reasoning work with geometry that occurs in the middle grades. It is important, however, to provide opportunities for students to develop definitions out of their experiences rather than start with an abstract definition.

Transformations, particularly on a coordinate grid, enable connections to be made between geometry and algebra as students operate with coordinates and transformation rules (e.g., adding −3 to every x-coordinate and 5 to every y-coordinate) or between geometry and measurement as students compare the angle measures or lengths of a figure and its image under a translation, rotation, reflection, or size change. We want students to make connections and realize how concepts studied in one context connect to concepts in other contexts so that they develop flexibility in their mathematical understanding.

Lesson: *Developing Definitions for Geometric Figures*

The following lesson is designed to help students develop a mathematical understanding of the definition of a specific quadrilateral that they likely have not studied in elementary school, namely a kite.

TEACHER: Today we're going to explore a geometric figure that you may not have studied before, but that you probably already know something about. Have any of you every flown a kite? (*One or two students raise their hands.*) Even though a lot of you haven't flown a kite, what do you know about kites?
STUDENT: People fly them in the air.
STUDENT: They come in all kinds of shapes and colors.
STUDENT: People often fly them near water. You need a place with a lot of wind.

TEACHER: Those are all good observations. I want you to look at this table I have displayed on the overhead. It lists some figures that are kites and some that are not kites (see Figure 7–7). I want you to talk at your table about what you might list as some properties or characteristics of a kite. We want to see if we can determine a definition. (*Students had about ten minutes to work together before the teacher brought them back for a whole-class discussion.*) Let's come back together and see if we can list some characteristics of the figures in the column labeled *kites*. We can also list characteristics of those figures in the *not kites* column to help us differentiate between the two. Someone want to start?

STUDENT: All the kites are quadrilaterals. Some of the *not kites* are not quadrilaterals—like the triangle.

STUDENT: In kites, there are pairs of congruent sides.

STUDENT: You have that in the *not kites* column too.

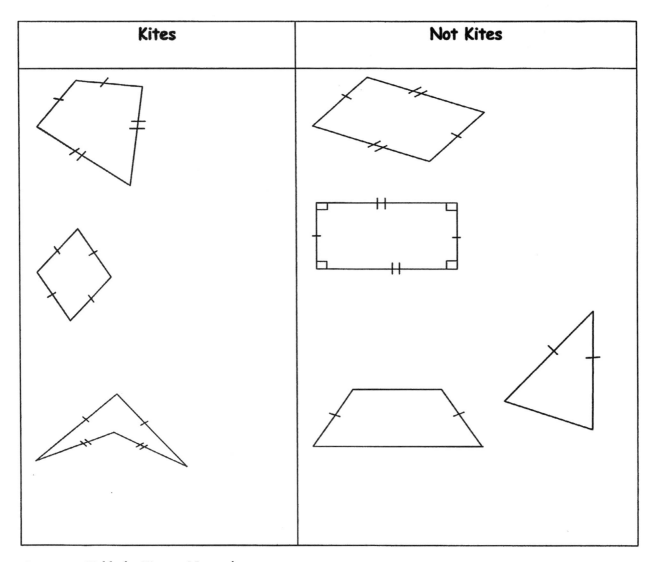

Figure 7–7 *Table for* Kites or Not *task.*

TEACHER: Let's stop here for a minute. Is it true that in the kites column we always have pairs of congruent sides? (*Students agree.*) But we also heard that we had pairs of congruent sides in the not kites column. So, how might we clarify this characteristic so it relates to kites but not to figures that aren't kites? (*Silence for a few seconds.*)

STUDENT: In the not kites column, the same marks in the figure are always on opposite sides. In the kites column, for two of the figures the same marks are on sides next to each other. I'm not quite sure what to do with this one. (*pointing to the rhombus*)

TEACHER: Let's hold off on your last statement for just one minute. Let's consider the other statements you made. He suggested that the congruent sides in a kite are next to each other and in a non-kite they are opposite each other. What do the rest of you think about that distinction?

STUDENT: I think it works. It even works with the rhombus, the one with the same markings on all four sides. If you take the top and bottom sides on the left they are next to each other, and they have the same markings. Then the top and bottom sides on the right are also next to each other and have the same markings. So I think his idea works.

TEACHER: Any other thoughts?

STUDENT: A rhombus seems like it could be in both columns. I get how the two sides next to each other are congruent but the opposite sides are also congruent.

TEACHER: So if we could only give one example of a kite, do you think a rhombus would be a good example?

STUDENT: No, because it seems to be a special case. It's like the kites, and it's also like the top two quadrilaterals in the not kites example.

TEACHER: That's a good way to think about it. So, do we have enough information to decide what a kite is?

STUDENT: A kite is a quadrilateral that has two pairs of sides congruent. The congruent pairs need to be next to each other.

TEACHER: A word you might not have heard before is *adjacent*, meaning next to each other and sharing a common vertex. So, we could state your definition as a quadrilateral with two pairs of adjacent sides congruent. I want to show you a figure and have you decide if you think this is a kite and whether or not it fits our definition. (See Figure 7–8.)

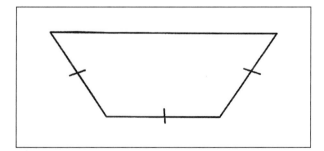

Figure 7–8 *A figure to test the definition of a kite.*

STUDENT: I don't think it's a kite because it looks like a trapezoid and we have a trapezoid in the not kites column. But you can have two pairs of adjacent sides congruent if you take the left side and the bottom and then the right side and the bottom. So it seems like it is a kite based on our definition. I think something's wrong.

TEACHER: You've raised an interesting dilemma for us. We have a figure that you don't think should be a kite, but it seems to fit the definition. So, what should we do?

STUDENT: I think we might need to fix our definition. Is it okay to use the bottom side twice when you make your congruent pairs? It doesn't seem like you should be able to use a side more than one time.

TEACHER: So you're suggesting a possible revision to the definition. Can you try to put your revision into the definition we had earlier for a kite?

STUDENT: A kite is a quadrilateral with two different pairs of adjacent sides congruent.

STUDENT: The left and bottom and right and bottom are two different pairs. So this doesn't solve the problem.

TEACHER: Well, the word *distinct* is often used to describe these pairs with the implication that the pairs cannot have a side in common. So, that helps solve the problem of using the bottom side in both pairs. Would you please restate the definition one more time using this idea of distinct?

STUDENT: A kite is a quadrilateral with two distinct pairs of congruent adjacent sides.

In this lesson, the teacher engaged students in an exercise in which they developed their own definition for a class of quadrilaterals they had likely not studied before. The teacher helped students tease out characteristics that distinguish kites from other quadrilaterals and then helped them refine their definition more than one time, interjecting specialized language (*adjacent* and *distinct*) only when that language was necessary to clarify the definitions.

We need to remember that definitions are often abstract for students. When we help them reason through similarities and differences among figures and develop their own definitions, we increase the likelihood that they will be able to describe that class of figures when they study it again.

Lesson: *Transformations: Reflections*

Students in the elementary grades likely explored reflections, translations, and rotations and learned that these movements result in images that are congruent to the preimage. They may have had some experience with size changes and learned that this transformation yields an image that is similar to the preimage. Now, students can use coordinates to describe transformations analytically and to justify relationships between the preimage and the image.

Developing spatial sense is similar to developing number sense. When students are provided multiple experiences with shapes and spatial relationships over time, their spatial sense is strengthened. These experiences create opportunities for students to reason spatially about shapes and relationships among shapes. Students' work in geometry should be one of doing. "As students sort, build, draw, model, trace, measure, and construct, their capacity to visualize geometric relationships will develop" (NCTM 2000, 165). These experiences strengthen spatial reasoning skills and promote students' abilities to make mathematical arguments to justify conjectures about geometric relationships.

Consider the following lesson, based on the CD task *Reflection, Reflection*, in which students work with figures on a coordinate grid to determine a symbolic way to represent reflections.

TEACHER: I have three sets of coordinates on the overhead for three different triangles. I want two different groups each to take one set of coordinates. I want you to plot the triangle, then reflect it over the x-axis, and then determine the coordinates of the image. Label your image points A', B', and so forth.

set 1: $A = (3, 2)$, $B = (5, 2)$, $C = (4, 5)$

set 2: $D = (-1, 0)$, $E = (-2, 3)$, $F = (-5, 1)$

set 3: $G = (-1, -1)$, $H = (1, -1)$, $I = (0, -2)$

I'm going to give each group a transparency with a grid so you can plot the triangle and the image and then share with the rest of the class. (*Students are given time to work in groups, with the recorder transferring the figures to the transparent grid.*) Let's see what you found. What are the coordinates of the image for each triangle?

set 1: $A' = (3, -2)$, $B' = (5, -2)$, $C' = (4, -5)$

set 2: $D' = (-1, 0)$, $E' = (-2, -3)$, $F' = (-5, -1)$

set 3: $G' = (-1, 1)$, $H' = (1, 1)$, $I' = (0, 2)$

Take a look at our results. What conjecture can you make about the coordinates (x, y) when you reflect over the x-axis?

STUDENT: The x-coordinate doesn't change. The y-coordinate changes sign; if it was positive it becomes negative and if it was negative it becomes positive.

TEACHER: How many of you have the same thing? (*There is general agreement around the room.*) How could we write this reflection using our mapping notation from before? (*Blank stares.*) It seems that I have a lot of puzzled faces in the room. Let me write the following:

$(x, y) \rightarrow$ _____

How can you complete the blank to describe what happens when you reflect over the x-axis?

STUDENT: You just get x and the opposite of y, so $(x, -y)$.

TEACHER: Does that make sense, and if so, why?

STUDENT: Yes, it does. When you reflect over the x-axis, you just move from above the x-axis to below or vice versa. So your x doesn't change because you just moved up or down. The y-coordinate tells how far you are above or below the x-axis. So when you move from above to below or below to above the sign has to change.

TEACHER: How many of you agree? Suppose I wanted you to reflect over the y-axis. I want you to make a conjecture about what the transformation would look like with coordinates, and then I want you to reflect your triangle over the y-axis to check whether your conjecture agrees with what you did.

In this lesson, students worked with coordinates to represent a geometric result using algebra. Students' prior work with coordinates made this task a relatively easy one. But the teacher focused students to reason about why the symbolic representation for the transformation made sense. Then she had them expand that reasoning by making and testing a conjecture about the result if students reflected over the y-axis instead of the x-axis. One advantage of having students work with transformations on a coordinate grid is that they are able to reinforce their algebra skills within a geometric context in which the visual result helps identify errors that might have occurred in the algebra.

A natural extension of the lesson on reflections would be to have students compare the lengths of the sides of the preimage and image triangles to determine that the figures are congruent. Even if students recognize that the preimage and image are congruent from the visual plots or from their work in elementary school, the coordinates provide an opportunity to generate an algebraic deductive proof for a geometric result.

Measurement

Students in grades 6 through 8 should realize that measurement is present throughout their everyday lives. Not only is measurement intertwined throughout the content standards, but it is also integrated in other disciplines. One important idea for students to understand about measurement is that measures are approximations and that we approximate measures daily. Understanding that approximations are acceptable and useful will help students when they are not able to use direct comparison to complete a measurement task.

Students are extending and deepening their understandings about the measurement of length, volume, weight, capacity, and time throughout the upper elementary and middle grades. They are exploring and building foundational understandings about perimeter and area. They are finding that it is important to be accurate in their measurements and that some tools measure more precisely than others. Students are using a variety of measurement tools and making decisions about which ones are more appropriate to use in a variety of measuring situations. We must insist that our students explain their reasoning in thinking about how to measure and why the measuring tool they've chosen is appropriate.

Allowing students to measure objects found in familiar contexts is critical to their development of measurement concepts and skills. Measuring objects from the classroom and the immediate environment is motivating and relevant for students. Students need to measure in the customary (English) as well as the metric system. Although we don't generally expect students to convert between systems, they should be encouraged to generate benchmarks that connect the two systems, such as that a meter is slightly longer than a yard, in order to help them make sense of measures in either system. The relationships within systems are used as students understand three feet equals a yard or a gallon equals four quarts. Although we might expect that middle grades students will be familiar with the customary system, their knowledge of the relationships within the system may be fragile. In the metric system, students should recognize that the units are related by factors of ten, such as ten centimeters equals a decimeter or one hundred centimeters equals a meter. Students need to realize the importance of including the unit of measure with the number when recording their measurements; a number without a related measurement unit does not convey the meaning it needs to convey.

Estimation in situations involving measurements should be included in our instruction. Students should relate the size of a unit to the quantity of the estimation and see that the unit size can help determine the reasonableness of an estimate. To support students in developing their estimation skills, facilitate student discussions that allow students to share their estimation strategies and to reason about the effectiveness of other strategies being shared.

Moving too quickly to students' use of formulas to complete measurement tasks is detrimental to their conceptual development of these ideas. Even in the middle grades, students benefit from instruction that allows them to develop understandings through hands-on experiences in which they focus on concepts prior to applying formulas. Too many students can verbalize a formula without understanding when that formula is appropriate and when it is not (think how often your students describe area as length times width).

Lesson: *Similar Figures and Areas*

Measurement related to similar figures provides numerous opportunities for students to reason about important relationships. Students often have misconceptions about the area of similar figures, believing that the area of a figure whose sides are twice as long as another figure is also twice as large. That is, students fail to distinguish linear relationships, such as perimeter, from relationships that are two-dimensional, such as area. Students need opportunities to build similar figures and to reason through these relationships on their own, as indicated in the following lesson, based on the CD task *Similar Figures and Area*.

TEACHER: We've worked with similar figures before when we've studied size changes. Today we're going to look at how the perimeters and areas of similar figures compare. Before we begin, I'd like someone to remind us what the term *scale factor* means when we're talking about size changes and similar figures.

STUDENT: It's the number that tells how to change the sides of a figure.

TEACHER: I need you to say a bit more. Change the sides how?

STUDENT: It's the number that you multiply each of the sides by to make the figure bigger or smaller.

TEACHER: Okay, so it's the multiplier for each side. Let's use that idea to draw some figures and then their images with different scale factors. I'm going to give each group a different rectangle and scale factor. I want you to sketch your figure and find its perimeter and area. Then I want you to sketch the image with the given scale factor and find the perimeter and area of the image. I'm going to put a table on the board and I want you to record your results in the table so we can compare them for different rectangles. (See Figure 7–9.)

(Students had some time to work together, with groups putting their results on the board. After all the results were posted on the table, the class came back together for a discussion about the results.) Let's look at the table of values. What can you say about the perimeters of the original figure and its image?

STUDENT: Three of them are bigger and one is smaller.

TEACHER: Okay, describe how much bigger or how much smaller.

STUDENT: The first one went up by 20, the next by 14, the next by 18, and the last one went down by 6.

TEACHER: That's certainly true. Can we see any pattern here that would help us make a generalization about how the perimeters compare?

STUDENT: I don't think we should just look at adding. I noticed that you could multiply the first perimeter by 3 and get 30, and then the next one by 2 and get 28, and then the next one by 4 and get 24. For the last one, if you divide 12 by 2 you get 6. That's the same as multiplying by half.

TEACHER: You seem to have hit on a possible idea here. Can you take what you've just said and make a conjecture about how the perimeters compare for the original figure and its image?

STUDENT: Well, the multiplier in each case is the scale factor. So my conjecture is that the perimeter of an image figure is the perimeter of the original figure times the scale factor.

TEACHER: Can anyone explain why that conjecture might make sense, or give us a counterexample?

STUDENT: Perimeter is the distance around a figure. So if you multiply every side by a number, then the whole distance gets multiplied by that number. I think the conjecture makes sense.

TEACHER: We have one possible explanation for the conjecture. Anybody else have a thought? Could anybody justify what was just said with symbols?

STUDENT: You could write the perimeter of the original rectangle as $l + w + l + w$. If s is the scale factor, then the new figure has length sl and width sw. So its perimeter is $sl + sw + sl + sw$. If I use the distributive property, I get $s(l + w + l + w)$, which is just the scale factor times the original perimeter. So it's always true.

TEACHER: That's a great deductive argument. So we have one justification that describes the result in words and one that verifies it with symbols. I'd like you to work together to discuss what conjectures can be written for how the areas com-

Original Dimensions	Perimeter	Area	Scale Factor	New Dimensions	New Perimeter	New Area
2 in. × 3 in.	10 in.	6 sq in.	3	6 × 9	30 in.	54 sq in.
3 in. × 4 in.	14 in.	12 sq in.	2	6 × 8	28 in.	48 sq in.
1 in. × 2 in.	6 in.	2 sq in.	4	4 × 8	24 in.	32 sq in.
2 in. × 4 in.	12 in.	8 sq in.	$\frac{1}{2}$	1 × 2	6 in.	2 sq in.

Figure 7–9 *Perimeter and Area results for similar rectangles.*

pare. (*Students are given time to work together to develop and verify a conjecture about how the areas of the preimage and image compare.*) Would someone like to share what their group found?

STUDENT: We first looked at the numbers in the table to see how they compare. Since we multiplied with perimeter, we decided to see if there was a way to multiply to get from the first area to the second. In the first row, we multiplied 6 times 9 to get 54, then 12 times 4 to get 48, then 2 times 16 to get 32, and 8 times one-fourth to get 2. We decided our conjecture is square the scale factor and multiply by the original area to get the new area.

TEACHER: I really like how you used ideas about the perimeter comparison to help you get started in making a comparison about the areas. Did the rest of you get the same type of conjecture? (*General agreement seems to be present.*) Can someone give us an argument that explains why this might make sense?

STUDENT: We used the idea from before with perimeter. The area of the original figure is lw. The area of the image is $(sl)(sw)$ which is s^2lw. So the conjecture is true.

In this lesson, students had an opportunity to build specific figures in order to explore these measurement concepts in a concrete manner. Also, students explored perimeter and area simultaneously so they could compare and contrast these two measures in similar figures. Notice that students have an opportunity in this lesson to integrate geometry, measurement, and algebra in addition to the process standards of problem solving, representation, and connections. Students demonstrated the importance of sharing their thinking with peers, as the strategies for comparing area relied on strategies discussed in the class about comparing perimeters.

Data Analysis and Probability

Throughout the middle grades, students build on their earlier understandings of how to represent, interpret, and analyze data. They represent data using tables, line plots, double-bar graphs, and line graphs. Interpretation and analysis of data are more complex than in the elementary grades, and students' reasoning about the data can be used to make predictions.

In grades 6 through 8, students look at data sets and compare one set of data with another. Students need to attend to the sizes of the data sets, perhaps using proportional reasoning to make comparisons among data sets with different sizes. That is, students need to use multiplicative reasoning as the basis of their comparisons rather than additive reasoning. In addition to working with the visual aspects of data sets, students also work with measures, including measures of center (e.g., mean, median, mode) as well as measures of spread (e.g., range, interquartile range). As students participate in discussions about these varied ideas, encourage them to use mathematical language precisely to describe relationships in the data sets.

In addition to determining appropriate displays for specific types of data, students should be reasoning about the data, including formulating generalizations, making predictions, and developing mathematical arguments about the data they are analyzing. Students are capable of reasoning about whether a data set collected from a local source would be representative of a data set collected from another location or source, and they can reason about what is the cause for the differences.

Reasoning about probability is unique because students are reasoning within a context of uncertainty. In the elementary grades, students likely explored basic concepts about probability and used terms such as *unlikely*, *likely*, and *equally likely*. In the middle grades, students continue to connect these terms to everyday situations, such as predicting the possibility of rain or the likelihood of winning a high-stakes lottery. Students analyze data collected from an investigation to estimate the probability of an event and make inferences about the situation based on the data. Probability becomes a means to assess the fairness of a game or situation.

Lesson: *Making Decisions Based on Data*

Students are familiar with teams in which coaches must decide the team members who will represent the school in a competition. How can data be used to help make decisions about such choices? Often there are many decisions that can be made and many factors that influence decisions. Consider the *Choose That Swimmer* task:

Wade, Alexander, and Joshua are the top swimmers for their school team. The table below shows their placements in each of 6 races for their most recent swim meet.

Swim stroke	100 m backstroke	200 m butterfly	50 m freestyle	1500 m freestyle	200 m breast-stroke	200 m individual medley
Wade	1	4	2	6	3	2
Alexander	3	3	3	2	2	3
Joshua	2	1	4	5	4	1

For an upcoming competition, the school team can only send two swimmers. Based on these results, which two swimmers should the team send? Justify your decision.

TEACHER: I have data here from three swimmers based on how they've done in several recent races. For the next competition, the coach can only send two racers and wants our help to decide which swimmers to take. Think about this problem and determine a method to help the coach decide which two team members to take to the meet. (*Students are given an opportunity to work with the data.*) (See Figure 7–10.) Who would you recommend and why?

STUDENT: I started by finding the average of their places in all three races. (See Figure 7–11.) I decided that the lower the number, the better the swimmer was because that meant they placed higher in each race. So I would take Alexander and Joshua.

Figure 7–10 *Student working on the* Choose That
Swimmer *task.*

STUDENT: We decided to take Alexander and Joshua also, but we just added up their
numbers in the races. Alexander has 16 and Joshua has 17 so they have the two
lowest numbers. We didn't need to find the average.

STUDENT: We want to ask the coach more questions. Do the swimmers have to
swim in all six races?

TEACHER: Why would you be interested in knowing that information?

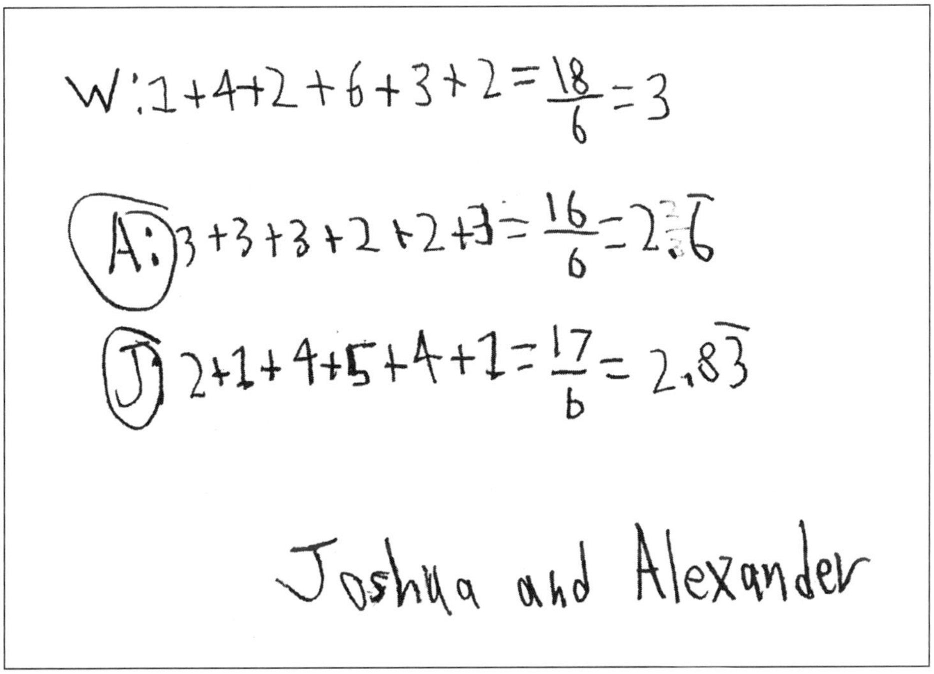

$$W: 1+4+2+6+3+2 = \frac{18}{6} = 3$$

$$A: 3+3+3+2+2+3 = \frac{16}{6} = 2.\overline{6}$$

$$J: 2+1+4+5+4+1 = \frac{17}{6} = 2.8\overline{3}$$

Joshua and Alexander

Figure 7–11 *Student's work to determine the two swimmers for the* Choose That
Swimmer *task.*

STUDENT: Well, if they have to swim all these races then we agree that the coach should choose Alexander and Joshua. But if they were only going to swim a couple of races you would need to look at the data for just those races. Like if you were only going to swim short races like the backstroke and the 50 m freestyle, then we would take Wade and Joshua because they do better at these races. So, it depends.

STUDENT: I would take Wade and Alexander. Wade gets a medal four times because he is in the top three for four races. Alexander gets a medal in every race, so he really needs to be on the team. Joshua only gets a medal three times. So if the school wants to win lots of medals, then I think the coach should take Wade and Alexander.

TEACHER: This is an interesting discussion. We've seen a couple of ways to make a decision here. We've also raised the idea that decisions are not always easily completed. In fact, we have some different recommendations to make to the coach, and we can justify those recommendations. There is often not just one way to make a decision, but data can help us make informed decisions.

This lesson highlights an important issue related to using data in making decisions. Different decisions are often possible with the same data sets as long as a justification can be made.

Lesson: *Making Predictions*

Concepts related to probability provide opportunities for students to make predictions and to consider differences between theoretical results and the results from an experiment. Consider the following dialogue based on the *Predict That Spin* task on the CD.

TEACHER: Consider the spinner I have on the overhead here. (See Figure 7–12.) I'd like you to estimate the fraction of the circle represented by each color on the spinner.

STUDENT: Green and purple are both half of a half of the circle, so each of those colors is $\frac{1}{4}$ of the circle. Black looks like it's about $\frac{1}{3}$. Since black and blue have to be half of the circle, that means blue is $\frac{1}{6}$.

TEACHER: I really like how you used what you know about parts of a circle to make sure that black and blue represent half. Suppose Charles has time on his hands and spins the spinner 200 times. He says he gets 60 green, 80 purple, 15 blue, and 45 black. Cody thinks Charles must have done something wrong to get these results. What do you think? Think about it for a couple of minutes on your own and then share with a partner.

STUDENT: Cody probably thought Charles was wrong because he thought the results would match up with his predictions. If we use the fractions from before, Charles should get 50 green, 50 purple, about 67 black, and 33 blue. So he didn't get the predictions. But that doesn't mean that Cody is right. Results can be way off even if the user doesn't do anything wrong.

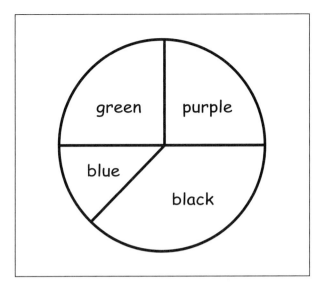

Figure 7–12 *Spinner for the* Predict That Spin *task.*

During this brief dialogue, students made connections to various ideas about fractions as they reasoned about the size of the four sectors of the spinner. More importantly, they recognized that theoretical probability does not have to match with the actual results, even if a spinner is fair. Students realized that the results from the second spin have nothing to do with the results of the first spin (i.e., the spins are independent). Although the theoretical results give us some idea of what is typical and what would be expected over the long run, unusual results (even if unlikely) do occasionally occur.

Final Thoughts

The interconnectedness of the mathematics process and content standards is demonstrated throughout this chapter in the student dialogues. When taught in this manner, students do not view mathematics as an isolated subject that has little relevance to them. Instead, they understand that the mathematics they are learning can be applied to other disciplines, to other mathematical ideas, and to real-world situations. They will be internalizing important ideas about mathematics as they engage in problem-solving tasks that promote reasoning and as they communicate and represent their reasoning in different ways. By presenting experiences that encourage our students to reason daily, we are enabling them to see that reasoning is relevant in all areas of their lives. We are also helping them realize that mathematics class is a place where discourse and conversation are expected.

Questions for Discussion

1. Reasoning takes a variety of forms across the content standards. In geometry, spatial reasoning is often not a topic explored in textbooks. Why is spatial

reasoning important? Discuss with colleagues several ideas for lessons that will develop students' spatial reasoning.

2. Algebra is frequently referred to as the gatekeeper to higher mathematics learning. How will you encourage your students to think algebraically? Do you see algebraic reasoning as something that can be done in all areas of mathematics? Why do you feel this way?

3. Reasoning about important mathematical ideas and relationships helps students make sense of mathematics and relate it to other ideas they will learn in high school. Think about all of the content areas: Identify a few important ideas that you can use as the basis for your students to reason and formulate conjectures. How will formulating generalizations about these ideas and developing mathematical arguments to justify them support your students in later grades?

If students are consistently expected to explore, question, conjecture, and justify their ideas, they learn that mathematics should make sense, rather than believing that mathematics is a set of arbitrary rules and formulas.

—Robert E. Reys, Mary M. Lindquist, Diana V. Lambdin, Nancy L. Smith, and Marilyn N. Suydam, *Helping Children Learn Mathematics*

Reasoning is viewed as a necessary process to ensure that students understand mathematical ideas and relationships. It is a way to make the abstract nature of mathematics understandable. Understandings about mathematics are further enhanced by an instructional program that presents knowledge that is connected as well as integrated and that helps students to think and reason about new knowledge by connecting their reasoning to what they already know.

It can be challenging to focus our instruction so that students become efficient problem solvers who reason effectively to make sense of the mathematics they are learning. We hope that this book will assist you in implementing this important change in your instructional program, a change that will greatly benefit your students for the rest of their lives. Students' ability to reason develops over time. The problems you select for your students to solve and the questions you ask of students in your daily discussions will help them learn how to select and use different types of reasoning, such as intuitive, inductive, and deductive thinking.

Through the establishment of a positive classroom environment, you are setting the stage for students' productive reasoning to occur. An environment in which students can feel safe and comfortable as they collaboratively share how they are reasoning supports them in making conjectures and developing mathematical arguments.

By providing your students with problems that are engaging, challenging, and authentic, with relevant connections, you will enable them to strengthen their reasoning skills and acquire mathematics knowledge through problem solving. These problems will promote rich discussions that will support your students' reasoning abilities.

Through your facilitation of these daily student discussions, you are enabling your students to become thoughtful thinkers as they reflect and clarify their thinking before expressing reasoning that is clearly understood. Your students will have multiple opportunities to listen to their classmates' thinking and reasoning, which enhances their own learning. When we make students' views the focus of our discussions, students are encouraged to listen to one another's reasoning and to question each other when reasoning needs clarification. Not only does this approach support students' reasoning abilities, but it also enables them to have a confident attitude about mathematics.

Through developing a repertoire of carefully chosen questions, you will enable your students to focus on effective reasoning about a variety of mathematical ideas and relationships. Asking questions that cause your students to ponder about mathematical ideas encourages reasoning and facilitates productive learning. Model for your students that they should also question and reason about what others have shared regarding the mathematics they are learning. You will be helping your students to ask questions of themselves and others, which allows them to monitor their own learning and become independent and powerful mathematical thinkers.

Through their participation in a variety of engaging activities provided on the CD, your students will develop, extend, and refine their reasoning skills. These activities provide you with evidence of their developing reasoning skills and mathematics knowledge. The many reasoning tools demonstrated in the book and on the CD will help your students learn to organize their thinking and reasoning so that they become independent problem solvers. The convenience and ease with which you can modify the CD activities to meet your students' needs, interests, and skills will certainly enhance your instruction.

The classroom environment we create to support our students' reasoning skills should be one that is integrated with the remaining process standards of problem solving, communication, representation, and connections. As you plan your instruction each school year, include daily instruction that infuses the process standards to help your students become proficient mathematical thinkers. Providing opportunities for your students to make connections among mathematical ideas, to other disciplines, and to their everyday lives will greatly enhance their reasoning abilities.

An important goal we share for all of our students is to ensure that they acquire critical reasoning skills that will serve them as they grapple with more complicated mathematics topics in later grades. As you implement new expectations into your classroom routine, you will encounter shared reasoning that contains flaws or misconceptions. We must convey to our students that flaws can occur in anyone's reasoning, and that the follow-up investigation to make sense of the misconception is indeed valuable to everyone's learning.

When our students know that their thinking and reasoning are a valued and necessary part of their development as mathematical thinkers, we will have indeed accomplished an important goal. As mathematics educators we want all of our students to understand that reasoning is a central part of every mathematics lesson and that they are responsible for expressing their reasoning and understanding the reasoning of others. This ability to reason helps our students make sense of the mathematics they are learning and enables them to become students who are willing to explore, investigate, and challenge mathematical ideas. Not only will our students benefit in amazing ways if we work with this goal in mind, but we will benefit also.

Additional Resources for Reasoning and Proof

The following resources are meant to support you as you continue to explore the reasoning and proof standard in grades 6 through 8. You will find a variety of text resources—articles that will provide you with instructional strategies or additional problem-solving activities that also promote reasoning. A list of mathematics websites is included to supply you with problem tasks, electronic manipulative ideas, and teacher resources.

Text Resources

Andre, R., and L. R. Wiest. 2007. "Using Sorting Networks for Skill Building and Reasoning." *Mathematics Teaching in the Middle School* 12 (February): 308–11.

Artzt, A. F., and S. Yaloz-Femia. 1999. "Mathematical Reasoning During Small-Group Problem Solving." In *Developing Mathematical Reasoning in Grades K–12*, Lee V. Stiff and Frances R. Curcio, eds., 115–26. Reston, VA: National Council of Teachers of Mathematics.

Aspinwall, L., and J. S. Aspinwall. 2003. "Investigating Mathematical Thinking Using Open Writing Prompts." *Mathematics Teaching in the Middle School* 8 (March): 350–53.

Austin, R. A., and D. R. Thompson. 1997. "Exploring Algebraic Patterns Through Literature." *Mathematics Teaching in the Middle School* 2 (February): 274–81.

Austin, R. A., D. R. Thompson, and C. E. Beckmann. 2003. "Exploring Measurement Concepts Through Literature: Natural Links Across Disciplines." In *Learning and Teaching Measurement*, Douglas H. Clements and George Bright, eds., 245–55. Reston, VA: National Council of Teachers of Mathematics.

Austin, R. A., D. R. Thompson, and C. E. Beckmann. 2004/2005. "Exploring Measurement Concepts Through Literature: Natural Links Across Disciplines." *Mathematics Teaching in the Middle School* 10 (January): 218–24.

Austin, R. A., D. R. Thompson, and C. E. Beckmann. 2006. "Locusts for Lunch: Connecting Mathematics, Science, and Literature." *Mathematics Teaching in the Middle School* 12 (November): 182–89.

Beckmann, C. E., D. R. Thompson, and R. A. Austin. 2004. "Exploring Proportional Reasoning Through Movies and Literature." *Mathematics Teaching in the Middle School 9* (January): 256–62.

Billings, E. M. H. 2001. "Problems That Encourage Proportion Sense." *Mathematics Teaching in the Middle School 7* (September): 10–14.

Bishop, J. W., A. D. Otto, and C. A. Lubinski. 2001. "Promoting Algebraic Reasoning Using Students' Thinking." *Mathematics Teaching in the Middle School 6* (May): 508–14.

Boats, J. J., N. K. Dwyer, S. Laing, and M. P. Fratella. 2003. "Geometric Conjectures: The Importance of Counterexamples." *Mathematics Teaching in the Middle School 9* (December): 210–15.

Bright, G. W., W. Brewer, K. McCain, and E. S. Mooney. 2003. *Navigating through Data Analysis in Grades 6–8*. Reston, VA: National Council of Teachers of Mathematics.

Bright, G. W., D. Frierson, Jr., J. E. Tarr, and C. Thomas. 2003. *Navigating through Probability in Grades 6–8*. Reston, VA: National Council of Teachers of Mathematics.

Bright, G. W., P. L. Jordan, C. Malloy, and T. Watanabe. 2005. *Navigating through Measurement in Grades 6–8*. Reston, VA: National Council of Teachers of Mathematics.

Bright, G. W., J. M. Joyner, and C. Wallis. 2003. "Assessing Proportional Thinking." *Mathematics Teaching in the Middle School 9* (November): 166–72.

Burrill, G. F. and P. C. Elliott. (Eds.) 2006. *Thinking and Reasoning with Data and Chance* (Sixty-eighth yearbook). Reston, VA: National Council of Teachers of Mathematics.

Carroll, W. M. 1999. "Using Short Questions to Develop and Assess Reasoning." In *Developing Mathematical Reasoning in Grades K–12*, Lee V. Stiff and Frances R. Curcio, eds., 247–55. Reston, VA: National Council of Teachers of Mathematics.

Chapin, S. H., and N. C. Anderson. 2003. "Crossing the Bridge to Formal Proportional Reasoning." *Mathematics Teaching in the Middle School 8* (April): 420–25.

Chappell, M. F., and D. R. Thompson. 2000. "*A Raisin in the Sun*: Fostering Cultural Connections in Mathematics with a Classic Movie." *Mathematics Teaching in the Middle School 6* (December): 222–225, 233–235.

Chappell, M. F., and D. R. Thompson. 2000. "Fostering Multicultural Connections in Mathematics through Media." In *Changing the Faces of Mathematics: Perspectives on African Americans*, Marilyn E. Strutchens, Martin L. Johnson, and William F. Tate, eds., 135–50. Reston, VA: National Council of Teachers of Mathematics.

Driscoll, M., and J. Moyer. 2001. "Using Students' Work as a Lens on Algebraic Thinking." *Mathematics Teaching in the Middle School 6* (January): 282–87.

Friel, S. N., W. O'Connor, and J. D. Mamer. 2006. "More than 'Meanmedianmode' and a Bar Graph: What's Needed to Have a Statistical Conversation?" In *Thinking and Reasoning with Data and Chance*, Gail F. Burrill and Portia C. Elliott, eds., 117–38. Reston, VA: National Council of Teachers of Mathematics.

Friel, S., S. Rachlin, and D. Doyle. 2001. *Navigating through Algebra in Grades 6–8*. Reston, VA: National Council of Teachers of Mathematics.

Gay, A. S., and C. J. Keith. 2002. "Reasoning about Linear Equations." *Mathematics Teaching in the Middle School 8* (November): 146–48.

Goldsby, D. S., and B. Cozza. 2002. "Writing Samples to Understand Mathematical Thinking." *Mathematics Teaching in the Middle School 7* (May): 517–20.

Greenes, C., and C. Findell. 1999. "Developing Students' Algebraic Reasoning Abilities." In *Developing Mathematical Reasoning in Grades K–12*, Lee V. Stiff and Frances R. Curcio, eds., 127–37. Reston, VA: National Council of Teachers of Mathematics.

de Groot, C. 2001. "From Description to Proof." *Mathematics Teaching in the Middle School 7* (December): 244–48.

Hawes, K. 2006/2007. "Using Error Analysis to Teach Equation Solving." *Mathematics Teaching in the Middle School 12* (December/January): 238–42.

Jones, G. A., C. A. Thornton, C. W. Langrall, and J. E. Tarr. 1999. "Understanding Students' Probabilistic Reasoning." In *Developing Mathematical Reasoning in Grades K–12*, Lee V. Stiff and Frances R. Curcio, eds., 146–55. Reston, VA: National Council of Teachers of Mathematics.

Kim, O., and L. Kasmer. 2007. "Using 'Prediction' to Promote Mathematical Reasoning." *Mathematics Teaching in the Middle School* 12 (February): 294–99.

Krebs, A. 2005. "Studying Students' Reasoning in Writing Generalizations." *Mathematics Teaching in the Middle School* 10 (February): 284–87.

Krulik, S., and J. A. Rudnick. 1999. "Innovative Tasks to Improve Critical- and Creative-Thinking Skills." In *Developing Mathematical Reasoning in Grades K–12*, Lee V. Stiff and Frances R. Curcio, eds., 138–45. Reston, VA: National Council of Teachers of Mathematics.

Langrall, C. W., and J. Swafford. 2000. "Three Balloons for Two Dollars: Developing Proportional Reasoning." *Mathematics Teaching in the Middle School* 6 (December): 254–61.

Lanius, C. S., and S. E. Williams. 2003. "Proportionality: A Unifying Theme for the Middle Grades." *Mathematics Teaching in the Middle School* 8 (April): 392–96.

Lannin, J. K. 2003. "Developing Algebraic Reasoning Through Generalization." *Mathematics Teaching in the Middle School* 8 (March): 342–48.

Lannin, J., D. Barker, and B. Townsend. 2006. "Why, Why Should I Justify?" *Mathematics Teaching in the Middle School* 11 (May): 438–43.

Manouchehri, A., and D. A. Lapp. 2003. "Unveiling Student Understanding: The Role of Questioning in Instruction." *Mathematics Teacher* 96 (November): 562–66.

Martinie, S. L., and J. M. Bay-Williams. 2003. "Investigating Students' Conceptual Understanding of Decimal Fractions Using Multiple Representations." *Mathematics Teaching in the Middle School* 8 (January): 244–47.

Martinie, S. L., and J. M. Bay-Williams. 2003. "Using Literature to Engage Students in Proportional Reasoning." *Mathematics Teaching in the Middle School* 9 (November): 142–48.

Moss, L. J., and B. W. Grover. 2006/2007. "Not Just for Computation: Basic Calculators Can Advance the Process Standards." *Mathematics Teaching in the Middle School* 12 (December/January): 266–71 plus lab sheets.

Parker, M. 2004. "Reasoning and Working Proportionally with Percent." *Mathematics Teaching in the Middle School* 9 (February): 326–30.

Pugalee, D. K., J. Frykholm, A. Johnson, H. Slovin, C. Malloy, and R. Preston. 2002. *Navigating through Geometry in Grades 6–8*. Reston, VA: National Council of Teachers of Mathematics.

Rachlin, S., K. Cramer, C. Finseth, L. C. Foreman, D. Geary, S. Leavitt, and M. S. Smith. 2006. *Navigating through Number and Operations in Grades 6–8*. Reston, VA: National Council of Teachers of Mathematics.

Rubenstein, R. N. 2002. "Building Explicit and Recursive Forms of Patterns with the Function Game." *Mathematics Teaching in the Middle School* 7 (April): 426–31.

Shaughnessy, J. M. 2006. "Research on Students' Understanding of Some Big Concepts in Statistics." In *Thinking and Reasoning with Data and Chance*, Gail F. Burrill and Portia C. Elliott, eds., 77–98. Reston, VA: National Council of Teachers of Mathematics.

Smith, M. S. 2001. *Practice-Based Professional Development for Teachers of Mathematics*. Reston, VA: National Council of Teachers of Mathematics.

Smith, M. S., E. A. Silver, and M. K. Stein with M. Boston, M. A. Henningsen, and A. F. Hillen. 2005. *Improving Instruction in Rational Numbers and Proportionality: Using Cases to Transform Mathematics Teaching and Learning, Volume 1*. New York: Teachers College Press.

Smith, M. S., E. A. Silver, and M. K. Stein with M. A. Henningsen, M. Boston, and E. K. Hughes. 2005. *Improving Instruction in Algebra: Using Cases to Transform Mathematics Teaching and Learning, Volume 2*. New York: Teachers College Press.

Smith, M. S., E. A. Silver, and M. K. Stein with M. Boston and M. A. Henningsen. 2005. *Improving Instruction in Geometry and Measurement: Using Cases to Transform Mathematics Teaching and Learning, Volume 3*. New York: Teachers College Press.

Sowder, L., and G. Harel. 1998. "Types of Students' Justifications." *Mathematics Teacher* 91 (November): 670–75.

Stein, M. K. 2001. "Putting Umph into Classroom Discussions." *Mathematics Teaching in the Middle School* 7 (October): 110–12.

Stein, M. K., M. S. Smith, M. A. Henningsen, and E. A. Silver. 2000. *Implementing Standards-Based Mathematics Instruction.* Reston, VA: National Council of Teachers of Mathematics and New York: Teachers College Press.

Tarr, J. E., H. S. Lee, and R. L. Rider. 2006. "When Data and Chance Collide: Drawing Inferences from Empirical Data." In *Thinking and Reasoning with Data and Chance,* Gail F. Burrill and Portia C. Elliott, eds., 139–50. Reston, VA: National Council of Teachers of Mathematics.

Thompson, C. S., and W. S. Bush. 2003. "Improving Middle School Teachers' Reasoning about Proportional Reasoning." *Mathematics Teaching in the Middle School* 8 (April): 398–403.

Thompson, D. R., and R. A. Austin. 2004. "Socrates and the Three Little Pigs: Connecting Patterns, Counting Trees, and Probability," including teaching notes. In *Exploring Mathematics Through Literature: Articles and Lessons from Prekindergarten Through Grade 8,* Diane Thiessen, ed., 217–29. Reston, VA: National Council of Teachers of Mathematics.

Thompson, D. R., M. T. Battista, S. Mayberry, K. Y. Yeatts, and J. S. Zawojewski. (in press). *Navigating through Problem-Solving and Reasoning in Grade 6.* Reston, VA: National Council of Teachers of Mathematics.

Thompson, D. R., R. A. Austin, and C. E. Beckmann. 2002. "Using Literature as a Vehicle to Explore Proportional Reasoning." In *Making Sense of Fractions, Ratios, and Proportions,* Bonnie Litwiller and George Bright, eds., 130–37. Reston, VA: National Council of Teachers of Mathematics.

Thompson, D. R., M. F. Chappell, and R. A. Austin. 1999. "Exploring Mathematics through Asian Folktales." In *Changing the Faces of Mathematics: Perspectives on Asian Americans and Pacific Islanders,* Carol A. Edwards, ed., 1–11. Reston, VA: National Council of Teachers of Mathematics.

Thompson, D. R., M. F. Chappell, and R. A. Austin. 2002. "Using Native American Legends to Explore Mathematics." In *Changing the Faces of Mathematics: Perspectives on Indigenous People of North America,* Judith Elaine Hankes and Gerald R. Fast, eds., 129–43. Reston, VA: National Council of Teachers of Mathematics.

Tsuruda, G. 1994. *Putting It Together: Middle School Math in Transition.* Portsmouth, NH: Heinemann.

van Dooren, W., D. de Bock, L. Verschaffel, and D. Janssens. 2003. "Improper Applications of Proportional Reasoning." *Mathematics Teaching in the Middle School* 9 (December): 204–9.

Watson, J. M., and J. M. Shaughnessy. 2004. "Proportional Reasoning: Lessons from Research in Data and Chance." *Mathematics Teaching in the Middle School* 10 (September): 104–9.

Web Resources

The following websites provide a variety of lesson ideas, classroom resources, and ready-to-use problem-solving and reasoning tasks.

> **www.eduplace.com/math/brain/index.html**—This Houghton Mifflin website, called Education Place, contains brain teasers for grades 3 through 8 as well as an archive of past problems.

> **www.etacuisenaire.com**—This website is run by ETA/Cuisenaire, a supplier of classroom mathematics manipulatives and teacher resource materials.

www.heinemann.com—This website of Heinemann Publishing offers a variety of professional development resources for teachers.

illuminations.nctm.org/—On this National Council of Teachers of Mathematics (NCTM) website you will find interactive standards-based activities for students. Standards-based lessons are also included that promote problem solving and reasoning.

mathforum.org/kb/forumindex.jspa—This site offers a list of discussion forums by subject area. Teachers post problems or discussion topics so that others can reply with solutions or suggestions.

mathforum.org/mathmagic/—This website contains MathMagic posted challenges intended to motivate students to use computer technology while increasing problem-solving strategies and communication skills.

www.nctm.org—On this NCTM website you will find information on regional and national conferences sponsored by NCTM, as well as a variety of professional development materials.

nlvm.usu.edu/en/nav/vlibrary.html—This site offers most manipulatives in a virtual format.

www.whitehouse.gov/kids/math/—This site offers math challenges at elementary and middle school levels.

Baroody, A. J., and R. T. Coslick. 1998. *Fostering Children's Mathematical Power: An Investigative Approach to K–8 Mathematics Instruction.* Mahwah, NJ: Erlbaum.

Beckmann, C. E., D. R. Thompson, and R. A. Austin. 2004. "Exploring Proportional Reasoning Through Movies and Literature." *Mathematics Teaching in the Middle School 9* (January): 256–62.

Boats, J. J., N. K. Dwyer, S. Laing, and M. P. Fratella. 2003. "Geometric Conjectures: The Importance of Counterexamples." *Mathematics Teaching in the Middle School 9* (December): 210–215.

Burke, M. J. 1984. *The Use of Counterexample Logic by Adolescents.* Ph.D. dissertation, University of Wisconsin-Madison.

Burns, M. 1997. "How I Boost My Students' Number Sense." *Instructor* (April): 49–54.

Chapin, S. H., C. O'Connor, and N. C. Anderson. 2003. *Classroom Discussions: Using Math Talk to Help Students Learn, Grades 1–6.* Sausalito, CA: Math Solutions.

Chappell, M. F., and D. R. Thompson. 1999a. "Modifying Our Questions to Assess Students' Thinking." *Mathematics Teaching in the Middle School 4* (April): 470–74.

Chappell, M. F., and D. R. Thompson. 1999b. "Perimeter or Area? Which Measure Is It?" *Mathematics Teaching in the Middle School 5* (September): 20–23.

Countryman, J. 1992. *Writing to Learn Mathematics: Strategies that Work K–12.* Portsmouth, NH: Heinemann.

Crowley, M. L. 1987. "The van Hiele Model of the Development of Geometric Thought." In *Learning and Teaching Geometry, K–12*, Mary Montgomery Lindquist and Albert P. Shulte, eds., 1–16. Reston, VA: National Council of Teachers of Mathematics.

Daniels, H., and S. Zemelman. 2004. *Subjects Matter: Every Teacher's Guide to Content-Area Reading.* Portsmouth, NH: Heinemann.

Fraivillig, J. 2001. "Strategies for Advancing Children's Mathematical Thinking." *Teaching Children Mathematics 7* (April): 454–59.

Galbraith, P. L. 1981. "Aspects of Proving a Clinical Investigation of Process." *Educational Studies in Mathematics 12*: 1–28.

Kallick, B., and R. Brewer. 1997. *How to Assess Problem-Solving Skills in Math.* New York: Scholastic Professional.

Lannin, J., D. Barker, and B. Townsend. 2006. "Why, Why Should I Justify?" *Mathematics Teaching in the Middle School 11* (May): 438–43.

Lederhouse, J. N. 2003. "The Power of One-on-One." *Educational Leadership 60* (7): 69–71.

Lubienski, S. T., R. McGraw, and M. E. Strutchens. 2004. "NAEP Findings Regarding Gender: Mathematics Achievement, Student Affect, and Learning Practices." In *Results and Interpretations of the 1990–2000 Mathematics Assessments of the National Assessment of Educational Progress*, Peter Kloosterman and Frank K. Lester, Jr., eds., 305–336. Reston, VA: National Council of Teachers of Mathematics.

Manouchehri, A., and D. A. Lapp. 2003. "Unveiling Student Understanding: The Role of Questioning in Instruction." *Mathematics Teacher* 96 (November): 562–66.

McTighe, J., and F. T. Lyman, Jr. 1988. "Cueing Thinking in the Classroom: The Promise of Theory-Embedded Tools." *Educational Leadership* 45 (April): 18–24.

National Council of Teachers of Mathematics (NCTM). 2000. *Principles and Standards for School Mathematics*. Reston, VA: NCTM.

National Research Council. 2001. *Adding It Up: Helping Children Learn Mathematics*. Washington, DC: National Academy Press.

O'Daffer, P. G., and B. A. Thornquist. 1993. "Critical Thinking, Mathematical Reasoning, and Proof." In *Research Ideas for the Classroom: High School Mathematics*, Patricia S. Wilson, ed., 39–56. New York: Macmillan.

Reys, R. E., M. M. Lindquist, D. V. Lambdin, N. L. Smith, and M. N. Suydam. 2004. *Helping Children Learn Mathematics*. 7th ed. Hoboken, NJ: John Wiley and Sons.

Russell, S. J. 1999. "Mathematical Reasoning in the Elementary Grades." In *Developing Mathematical Reasoning in Grades K–12*, Lee V. Stiff and Frances R. Curcio, eds., 1–12. Reston, VA: National Council of Teachers of Mathematics.

Senk, S. L. 1985. "How Well Do Students Write Geometry Proofs?" *Mathematics Teacher* 78 (September): 448–56.

Sowder, L., and G. Harel. 1998. "Types of Students' Justifications." *Mathematics Teacher* 91 (November): 670–75.

Thompson, D. R., and S. L. Senk. 1993. "Assessing Reasoning and Proof in High School." In *Assessment in the Mathematics Classroom*, Norman L. Webb and Arthur F. Coxford, eds., 167–76. Reston, VA: National Council of Teachers of Mathematics.

Thompson, D. R., and S. L. Senk. 1998. "Using Rubrics in High School Mathematics Courses." *Mathematics Teacher* 91 (December): 786–93.

Valentine, C., T. P. Carpenter, and M. Pligge. 2005. "Developing Concepts of Justification and Proof in a Sixth Grade Classroom." In *Everyday Matters in Science and Mathematics: Studies of Complex Classroom Events*, Ricardo Nemirovsky, Ann S. Rosebery, Jesse Solomon, and Beth Warren, eds., 95–118. Mahwah, NJ: Erlbaum.

Van de Walle, J. A., and L. H. Lovin. 2006. *Teaching Student-Centered Mathematics, Grades 3–5*, Vol. 2. Boston: Pearson Education.

Yackel, E., and G. Hanna. 2003. "Reasoning and Proof." In *A Research Companion to Principles and Standards for School Mathematics*, Jeremy Kilpatrick, W. Gary Martin, and Deborah Schifter, eds., 227–36. Reston, VA: National Council of Teachers of Mathematics.

Why Are the Activities on a CD?

At first glance, the CD included with this book appears to be a collection of teaching tools and student activities, much like the activities that appear in many teacher resource books. But instead of taking a book to the copier to copy an activity, with the CD you can simply print off the desired page on your home or work computer. No more standing in line at the copier or struggling to position the book on the copier so you can make a clean copy. And with our busy schedules, we appreciate having activities that are classroom ready and aligned with our mathematics standards.

You may want to simplify some tasks or add complexity to others. When it is appropriate for your students, simply delete some sections (e.g., problems labeled *challenge*) for a quick way to simplify or shorten the tasks. This CD gives you much more than a mere set of activities. It gives you the power to create an unlimited array of problems that are suited to your students' interests, needs, and skills. Simply rename the file when saving it to preserve the integrity of the original activity. Here are some examples of ways you may want to change the tasks and why. A more complete version of this guide can be found on the CD.

Editing a Problem to Motivate and Engage Students

Personalizing Tasks or Capitalizing on Students' Interests

The editable CD provides a quick and easy way to personalize mathematics problems. Substituting students' names, the teacher's name, or a favorite restaurant, sports team, or location can immediately engage students. You know the interests of your students. Mentioning their interests in your problems is a great way to increase their enthusiasm for the activities. Think about their favorite activities and simply substitute their interests for those that appear in the problems.

For example, one teacher knew that many of her students participated on the school's swim team and decided to reword the *Choose That Swimmer* task to include names of some of her students as well as the name of the school's swim coach. (*Note:* This type of editing is also important when the problem situation may not be culturally appropriate for your students.)

Please advise no art in msp for About the CD-ROM Section, Please supply — comp.

Name _____

Choose That Swimmer

Wade, Alexander, and Joshua are the top swimmers for their school team. The table below shows the placements in each of six races for their most recent swim meet.

Swim stroke	100 m backstroke	200 m butterfly	50 m freestyle	1500 m freestyle	200 m breaststroke	200 m individual medley
Wade	1	4	2	6	3	2
Alexander	3	3	3	2	2	3
Joshua	2	1	4	5	4	1

For an upcoming competition, the school team can only send two swimmers. Based on these results, which two swimmers should the team send? Justify your decision.

_____

_____

_____

_____

_____

_____

_____

_____

_____

_____

_____

_____

_____

Name _____

Choose That Swimmer

Samantha, Lindsey, and Jennifer are the top female swimmers at Jefferson High School. The table below shows the placements in each of six races for their most recent swim meet.

Swim stroke	100 m backstroke	200 m butterfly	50 m freestyle	1500 m freestyle	200 m breaststroke	200 m individual medley
Samantha	1	4	2	6	3	2
Lindsey	3	3	3	2	2	3
Jennifer	2	1	4	5	4	1

For an upcoming competition, Jefferson High School can only send two swimmers. Based on these results, which two swimmers should the team send? Justify your decision.

_____

_____

_____

_____

_____

_____

_____

_____

_____

_____

_____

_____

_____

Editing a Problem to Differentiate Instruction

Scaffolding Tasks

Although we want all students to engage in complex tasks, some students may initially benefit from additional scaffolding or support as they begin learning to reason. Within a given class, some students may need additional support while others do not. By using the editable CD feature, you can add questions or hints to guide students who might need additional support. You can also have varied versions of a task for use by students within a class, so that you provide support only to those who need it and not to all students.

Name _____

Pyramid Ornaments

The *Marvelous Fun Summer Camp* decides to make square pyramid ornaments as decorations for a Fourth of July celebration. Each ornament has a square base with side length 14 centimeters and height 24 centimeters. The four triangular faces and the base will be decorated with patriotic designs.

Find the surface area to be decorated. Explain your reasoning to find the area.

_____
_____
_____
_____
_____
_____
_____
_____
_____
_____
_____
_____
_____
_____
_____
_____

Name _____

Pyramid Ornaments

The *Marvelous Fun Summer Camp* decides to make square pyramid ornaments as decorations for a Fourth of July celebration. Each ornament has a square base with side length 14 centimeters and height 24 centimeters. The four triangular faces and the base will be decorated with patriotic designs.

1. What is the area of the base? How do you know?

_____
_____

2. a. What information do you need to find the area of each triangular face?

_____
_____

   b. What information from (2a) do you already have? How can you find the missing information?

_____
_____

3. Find the surface area to be decorated. Explain your reasoning to find the area.

_____
_____
_____
_____
_____

Modifying Readability of Tasks

Adding some fun details can generate interest and excitement in real-life problems, but you might prefer to modify some problems for students with limited reading ability or for students with limited English. Simply deleting some of the words on the editable CD will result in an easy-to-read version of the same task so that students can still engage in the same mathematics without language proficiency being a stumbling block.

Name _____

Alike and Different

For each pair of figures, explain how they are alike and how they are different.

1. a square and a rectangle

2. a square and a rhombus

3. a rectangle and a parallelogram

4. a hexagon and an octagon

5. a circle and a sphere

Name _____

Alike and Different

For each pair of figures, explain how they are alike and how they are different.

1. a square ⬜ and a rectangle ▭

2. a square ⬜ and a rhombus ◇

3. a rectangle ▭ and a parallelogram ▱

4. a hexagon ⬡ and an octagon ⯃

5. a circle ◯ and a sphere ⊕

Modifying Data

Although all students may work on the same problem task, modifying the problem data will allow teachers to create varying versions of the task. Using the editable CD, you can either simplify the data or insert more challenging data, including larger numbers, fractions, decimals, percents, or variables.

Name _____

Dartboard Probabilities

Consider the dartboard at the right. The radii of the circles from smallest to largest are 0.75 inches, 1.5 inches, 2.25 inches, and 4.5 inches. The outside rectangle is 14 inches wide and 9.5 inches tall.

1. What is the probability of a dart hitting inside the smallest circle? Explain how you found your answer.

   _____

   _____

   _____

2. Number the circles from 1 (smallest) to 4 (largest). Find the probability of a dart hitting inside the ring between the second and third circles. Explain your thinking.

   _____

   _____

   _____

3. Sam said there is a greater probability of hitting the region on the dartboard outside the circle than inside the circle. Do you agree with Sam? Why or why not?

   _____

   _____

   _____

Name _____

Dartboard Probabilities

Consider the dartboard at the right. Each of the colored figures is a square. The lengths of the sides from smallest to largest are 2 inches, 4 inches, and 5 inches. The outside rectangle is 9 inches wide and 7 inches tall.

1. What is the probability of a dart hitting inside the smallest square? Explain how you found your answer.

   _____

   _____

   _____

2. Number the squares from 1 (smallest) to 3 (largest). Find the probability of a dart hitting inside the region between the second and third squares. Explain your thinking.

   _____

   _____

   _____

3. Sam said there is a greater probability of hitting the region on the dartboard outside the squares than inside. Do you agree with Sam? Why or why not?

   _____

   _____

   _____